Lecture Notes in Mechanical Engineering

About this Series

Lecture Notes in Mechanical Engineering (LNME) publishes the latest developments in Mechanical Engineering—quickly, informally and with high quality. Original research reported in proceedings and post-proceedings represents the core of LNME. Also considered for publication are monographs, contributed volumes and lecture notes of exceptionally high quality and interest. Volumes published in LNME embrace all aspects, subfields and new challenges of mechanical engineering. Topics in the series include:

- Engineering Design
- Machinery and Machine Elements
- Mechanical Structures and Stress Analysis
- Automotive Engineering
- Engine Technology
- Aerospace Technology and Astronautics
- Nanotechnology and Microengineering
- Control, Robotics, Mechatronics
- MEMS
- Theoretical and Applied Mechanics
- Dynamical Systems, Control
- Fluid Mechanics
- Engineering Thermodynamics, Heat and Mass Transfer
- Manufacturing
- Precision Engineering, Instrumentation, Measurement
- Materials Engineering
- Tribology and Surface Technology

More information about this series at http://www.springer.com/series/11236

M. Awang
Editor

2nd International Conference on Mechanical, Manufacturing and Process Plant Engineering

Springer

Editor
M. Awang
Mechanical Engineering Department
Universiti Teknologi Petronas
Seri Iskandar
Malaysia

ISSN 2195-4356 ISSN 2195-4364 (electronic)
Lecture Notes in Mechanical Engineering
ISBN 978-981-10-4231-7 ISBN 978-981-10-4232-4 (eBook)
DOI 10.1007/978-981-10-4232-4

Library of Congress Control Number: 2017935838

Printed on acid-free paper

This Springer imprint is published by Springer Nature
The registered company is Springer Nature Singapore Pte Ltd.
The registered company address is: 152 Beach Road, #21-01/04 Gateway East, Singapore 189721, Singapore

Contents

A Review of Friction Stir Welding Pin Profile

S. Emamian, M. Awang, F. Yusof, P. Hussain, M. Mehrpouya, S. Kakooei, M. Moayedfar and Adeel Zafar

Abstract Friction stir welding (FSW) is a solid state joining process where the joint occurs below the melting point of the base metal. Recently, not only soft materials but also high melting materials have been given an attention in FSW research activities. There are many studies on high melting materials such as steel, titanium, and inconel. There are different parameters which contribute to the joint quality. Many researchers have studied on the traverse and rotational speeds as the main process parameters in order to obtain optimum weld quality. Tool geometry is another important parameter that has an influence on the weld quality. This paper aims to review the development of various pin profiles and its effects to the microstructure and mechanical properties of the weld joint. Based on the published papers, square pin profile produced sound joints. However, in other studies threaded cylinder or threaded taper provides better joints. Above all, there is an equal result among all studies in which threaded shapes are most effective on tool performance.

Keywords FSW · Pin profile · Tool geometry · FSW parameters

1 Introduction

Friction Stir Welding (FSW) is a solid state joining process that was invented and patented by The Welding Institute (TWI) UK [1]. This invention is a big success for joining aluminum alloys as well as low melting temperature materials [2].

S. Emamian (✉) · M. Awang · P. Hussain · S. Kakooei · M. Moayedfar · A. Zafar
Department of Mechanical Engineering, Universiti Teknologi PETRONAS, 32610 Seri Iskandar, Malaysia
e-mail: sattar.emamian@gmail.com

F. Yusof
Department of Mechanical Engineering, University of Malaya, Kuala Lumpur, Malaysia

M. Mehrpouya
Department of Mechanical and Aerospace Engineering, Sapienza University of Rome, Rome, Italy

M. Awang (ed.), *2nd International Conference on Mechanical, Manufacturing and Process Plant Engineering*, Lecture Notes in Mechanical Engineering,
DOI 10.1007/978-981-10-4232-4_1

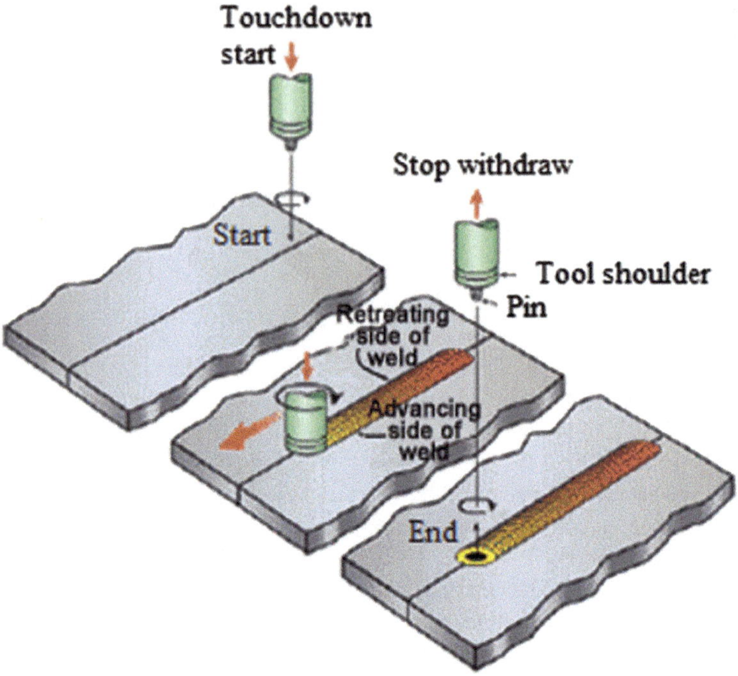

Fig. 1 Schematic of Friction Stir Welding [4]

Recently, FSW has been proven as an alternative joining technique for high melting materials such as steels, titanium, and so on. Figure 1 illustrates the schematic of FSW process. The applications of FSW can be found in marine, aerospace, automotive, rail and construction industries. Nowadays FSW is gradually being introduced to oil and gas industries such as offshore structure [3].

In FSW, the parameters that influence the quality of weld joint can be categorized into primary and secondary parameters. The primary parameters are traverse speed, rotational speed, and tool geometry. Meanwhile the secondary parameters are thickness of the workpiece, workpiece material, welding tool material, and pin profile.

In the last decade many researches have been done on various FSW parameters. Among them, there are some studies that concentrated on pin profiles. In order to obtain the best and optimum pin profiles, the geometry of pin will be evaluated with other parameters such as rotation and traverse speeds, and force.

Elangovan et al. [5–8] had many researches on pin profiles on aluminum alloys. They investigated the influence of pin profile and rotational speed, traversing speed, shoulder diameter, and axial force on the formation of friction stir processing zone. Furthermore, there are many other researchers who investigated the influence of pin profiles on microstructural and mechanical properties of the weld joint [9].

However, there are many challenging issues to optimize the welding process for joining hard materials. One of the issues is the appropriate tool pin profile design.

Tool geometry has an effect on weld joint for improving its properties and quality [10]. FSW tool has two main functions during the welding process. The tool shoulder rubs the workpiece and generates the heat and tool pin stirs the material [11]. In FSW, the tool materials also play very important role during the welding process for preventing the wear problem. The suitable tool material for welding a high melting workpiece should possess wear resistance with high strength as well as high toughness [12]. In addition, proper tool design reduces the force needed to push the tool along the weld line.

Based on the literatures, the popularity to study the pin profile is increased. However, there is little effort that focuses on FSW pin profiles. Therefore, this paper aims to review various pin profiles used in experimental, simulation and optimization works.

2 Experimental Studies on FSW Pin Profiles

Since the invention of FSW, there were many researches focused on developing the FSW process. Primarily, they considered low melting materials for base material such as Al, Mg, Zn. However, due to improvement of tool materials, the welding technique is also favorable for high melting materials such as Ti, Ni, and steel [2].

There are several studies concentrated on pin profile. Elangovan and Balasubramanian [5] investigated different pin profiles and their relation with rotational speeds. They selected straight cylindrical, tapered cylindrical, threaded cylindrical, triangular and square as the pin profiles. According to their results, square pin profile produced defect free joints regardless of rotational speeds. In addition, the result from 15 weld joints showed that the square pin profile with 1600 rpm rotational speed has superior tensile properties.

In another research, Elangovan and Balasubramanian [7] investigated the relationship between different pin profiles and tool shoulder diameter of AA6061 aluminum alloy welded joints. From their study, square pin profile with 18 mm shoulder diameter showed the higher tensile properties. They also investigated the effect of traveling speed for different pin profiles on the AA2219 weld strength. From their studies, square pin profiles showed defect free at the weld cross section. Moreover, of the three traveling speeds, 0.76 mm/s indicated better tensile properties regardless of tool pin profiles [6].

In another research Elangovan and Balasubramanian investigated the relation of different pin profiles with axial force. They fabricated the joints with three axial forces and five different pin profiles on AA6061 workpiece. They found that square pin profile with 7 kN showed superior tensile properties [8]. Figure 2 shows the schematic of the pin profile that was used in their studies.

In summary, it can be concluded that different FSW parameters has a significant effect on mechanical properties of the weldment.

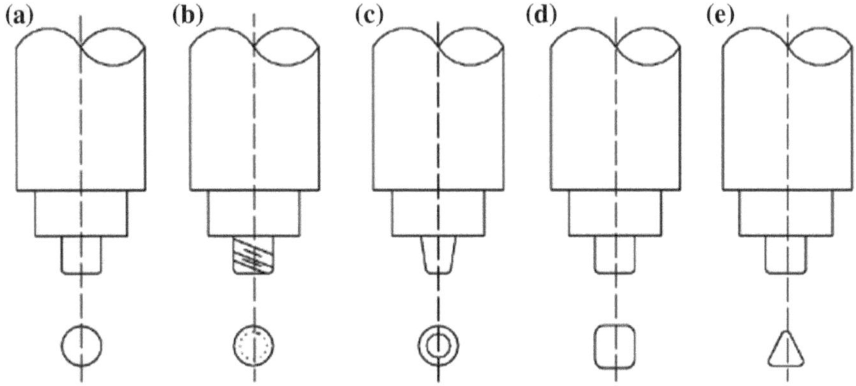

Fig. 2 FSW pin profiles. **a** Straight cylindrical. **b** Threaded cylindrical. **c** Tapered cylindrical. **d** Square. **e** Triangular [8]

Fig. 3 FSW pin profiles. **a** Threaded and tapered with three spiral flutes. **b** Threaded and tapered with triangle

Some experimental studies are carried out on different rotational speeds, traversing speeds in order to evaluate the relationship between pin profiles and FSW parameters. Patil and Soman [13] conducted an investigation on the influence of welding speed and the pin profiles of AA6082-O welded joint. They utilized tri-flutes and taper screw thread as the pin profiles with constant rotational speed of 1200 rpm and various welding speeds from 60 to 85 mm/min. The results indicated that pin profiles have significant role on mechanical properties and microstructure of the weld. Moreover, joints with tapered screw have higher tensile strength at a welding speed of 70 mm/min. It seems that the higher strength is because of the treaded type pin profile.

Xu et al. [14] used two different pin profiles to investigate the effect of rotational and traveling speeds on strength and ductility of 2219-T62 welded joint. Figure 3, shows the pin profiles that were used in the study. They concluded that, welding process parameters and pin profiles have a significant effect on strength and ductility. Furthermore, they found that the threaded and tapered with three spiral flutes showed higher mechanical properties result as compared to the other pin profile.

Ravindra and Surjya [15] studied the effect of pin profiles and the process parameters on the defect of AA5083 weld joint. Five different tool pin profiles (straight cylindrical, tapered cylindrical, triangular, square and cone) have been used to fabricate the joints at three different rotational speeds of 900, 1400 and 1800 rpm under a constant traverse speed of 16 mm/min. They found that the cylindrical profile produced defect free joints regardless of welding speeds. Moreover, cylindrical pin profile with high rotation speed produced defect free joints. However, the square and cylindrical profiles give better joints at 1400 rpm. On the other hand, at the low rotational speed of 900 rpm, triangular shapes produced sound welds. From the 15 joints that has been performed, they found that cylindrical pin profile at a rotational speed of 1400 rpm produced the best joint with highest tensile strength.

Salari et al. [16] studied the tool geometry and rotational speed on defect generation. In their work, they studied AA5456 aluminum alloy for the lap joints with two different thicknesses, 5 and 2.5 mm. In their study, they used 600 and 800 rpm for the rotational speeds with four different pin profiles (conical threaded, cylindrical–conical thread, stepped conical thread, and Neutral Flared Triflute). Schematic of pin profiles are illustrated in Fig. 4. The results showed significant influence of tool geometry on material flow as well as the mechanical properties. Furthermore, highest joint performance is achieved with the stepped conical threaded and highest tensile properties is obtained at rotational speed of 600 rpm irrespective of pin profiles.

Motalleb-Nejad et al. [17] studied the effect of pin profile on microstructure and mechanical properties of AZ31B magnesium alloy. Three pin profiles which are cylindrical, screw threaded cylindrical, and tapered were used in the study. The joints are welded at different traverse and rotational speeds. From the results they found that taper and screw threaded cylindrical produce defect free joints.

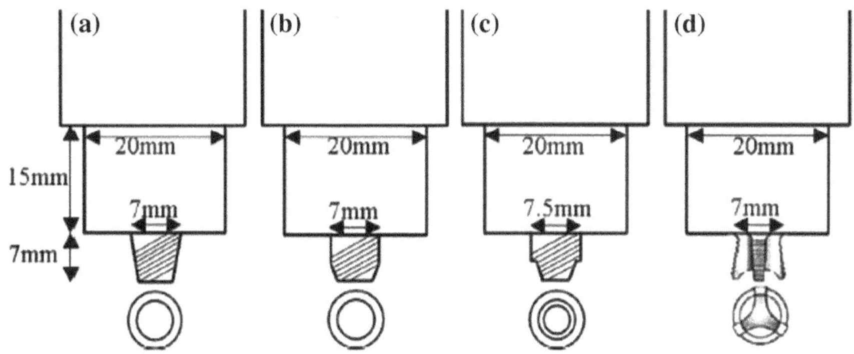

Fig. 4 Schematic of pin profiles. **a** Conical threaded. **b** Cylindrical-Conical threaded. **c** Stepped conical thread. **d** Neutral flared triflute

Moreover, they observed that rotational speed have higher effect on microstructure and mechanical properties. In addition, lower mechanical properties could be in the result of higher rotational speed and lower traveling speed. Consequently, the best mechanical properties were obtained with taper pin profile at $\frac{\omega^2}{\vartheta} = 6300$ where, ω is rotation speed and ϑ is the traverse speed.

Some other scholars tried to evaluate the effect of tool shapes with different welding speeds on mechanical properties. For instance, Trimble et al. [18] utilized the cylindrical, square and triflute for the FSW of 4.8 mm thick AA2024-T3 plates. The result showed that the triflute pin was the most effective shape for welding at high speeds. Figure 5 illustrates the schematic of three FSW pin profiles.

Trimble and his colleagues [19] in another study investigated the influence of pin profile on the microstructure and hardness of AA2024-T3 friction stir welded. They recommended to use pin profiles with flat surface or flutes over the tapered cylinder. Meanwhile, Ilangovan et al. [20] investigated the effect of pin profile on mechanical properties of dissimilar AA6061-AA5086 joints. In the study, of the three different pin profiles, threaded and tapered cylindrical produced defect free joints with the similar tensile properties. However, the threaded cylindrical is preferred due to its better joints shape. Figure 6 shows the schematic of the tool used in their study.

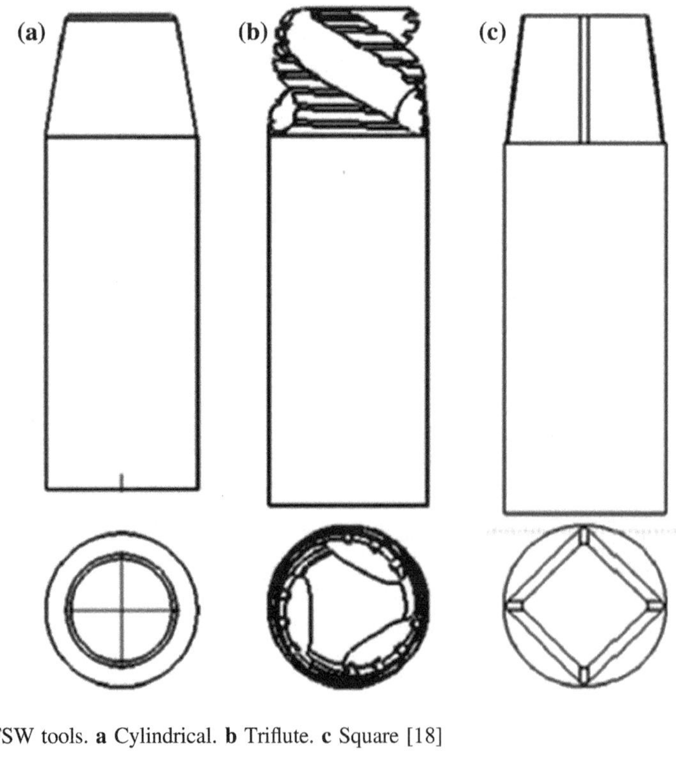

Fig. 5 FSW tools. **a** Cylindrical. **b** Triflute. **c** Square [18]

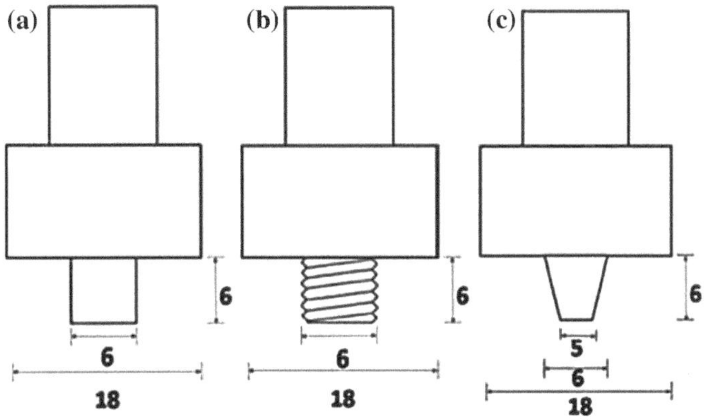

Fig. 6 FSW tools. **a** Cylindrical. **b** Threaded cylindrical. **c** Tapered [20]

Krasnowski et al. [21] investigated the influence of the pin profile on microstructure and mechanical properties of the Al6082 joint. They concluded that the shape of the FSW tool pin influences the mechanical properties.

Due to the significant effect of tool geometry on the material flow, fundamental correlation between material flow and microstructure of joints varies with each tool [11]. Nadikudi et al. [22] studied the formability of welds with different pin profiles. They found that square profile produced better formability properties as compared to other profiles. In addition, the weld microstructure made with a square profile had a fine grains with a uniformly distributed grains.

Some of the studies claim that rotational speed has not significant effect in the hardness profile of FSW welds. However, they mentioned that the rotational speed is influential on soften the material in the joints [23]. In addition, the above studies concluded that rotational and traversing speeds are effective on quality of the weld joints. Increasing the rotational speed and reducing the traversing speed will increase the temperature during FSW [24–26]. It is obvious that changing the geometry of the pin profile will influence on heat generation. Therefore, in order to obtain a defect free weld, the FSW speeds (rotational and traverse) and pin profiles must be chosen carefully.

Emamikhah et al. [27] considered the effect of pin profiles on FSW of high-zinc brass (CuZn40). They utilized chamfered taper, single-threaded taper, three-flute, threaded cylinder, threaded taper, spline, and hexahedron as the different pin profiles. Moreover, the rotation and traverse speeds were constant during the study. With embedding the thermocouples, heat generation is measured. The results revealed that threaded cylindrical pin profile have higher heat in advancing side, however, threaded taper have higher temperature and hardness on the retreating side. Consequently, higher heat will improve the material stirring below the shoulder and increase the mechanical properties. Nonetheless, the lower temperature was generated by the hexahedron profile. As mentioned earlier, the geometry of the pin can effect on the heat generation. Amount of friction between the tool and workpiece during stirring process may produce temperature. Therefore, the higher interaction, the higher temperature.

Some other studies considered the other parameters of FSW as compared to pin profiles. For instance, Mohanty et al. [28] examined the role of pin profile of FSW on aluminum and compared it with tool shoulder. They used chromium steel straight cylindrical, tapered cylindrical and trapezoidal for the experiment at 1400 rpm for rotation, 160 mm/min for welding speed and 3.5 kN for axial force. From the 27 joints experiment, they observed that straight cylindrical with 5 mm diameter have higher tensile strength.

Reza and Reynolds [29] studied the role of threaded pin on material flow and weldability of AA7050 and AA6061. They observed that using thread forms are very useful on tool performance and also reduces the force on tool.

Venkata Rao et al. [30] studied the microstructure and corrosion of AA 2219 Al-Cu alloy during FSW to evaluate the effect of pin profile. They pick out conical and triangular shapes for the experiment. From the results of electron backscattered diffraction and differential scanning calorimetric they found that corrosion resistance of triangular profile is better than conical. The results showed that heat generation is higher for triangular profile. With regards to the heat generation, the relative size of pin and shoulder is important in FSW process.

Venkata Rao et al. [31] studied the effect of pin profiles on microstructure of AA2219 alloy. They found the significant role of pin profiles in the microstructure of workpiece. The results showed that hexagonal tool produced higher temperature and uniform structure as compared to square, triangle, pentagon, and conical tool pin profiles.

In many experimental studies that the square pin profile is involved in, the sounds welds with higher mechanical properties produced by square shapes. On the other hand, threaded cylinder and threaded taper produced better joints in compare with other shapes. Once comparing different profiles in various studies, it can be summarized that threaded shapes are very influential on tool performance. According to aforementioned, some parameters such as workpiece material and thickness, rotation and traverse speeds have influence on joining quality. Therefore in order to select the optimum pin profiles, it depends on many parameters.

3 Simulation on FSW Pin Profiles

Not only the experimental works, but some researches also focused on simulation study. They tried to model different pin profiles and investigate the parameters such as heat generation, stress distribution, and effect of FSW parameters on mechanical properties and thermal characteristics. To model FSW process, a coupled thermo plastic behavior will be assumed in order to obtain the stress, strain as well as temperature distributions [32, 33]. There are many engineering software available to model FSW process. However, Kheireddine et al. [34] mentioned that for thermo-mechanical models, DEFORM-3D and ABAQUS software are appropriate to use.

Buffa et al. [35] used finite element model (FEM) to investigate the cylindrical and conical shapes with different angles. Thermo-mechanically coupled analysis has been performed to predict the process parameters. They observed that increasing the pin angle will enlarge the heat affected zone and thermal mechanical zone due to bigger weld nugget. In addition, plastic deformation would be increased. Due to helical movement of conical profile, material flow will also be increased.

Jaimin and Patil [36] investigated the mechanical properties and thermal analysis by modelling the different pin profiles for FSW on AA6061. Finite Element Analysis (FEA) has been used in order to modelling. The simulation was carried out at constant rotational speed of 600 rpm with different pin profiles; conical, hexagonal, trapezoidal, cylindrical flutes. They found that hexagonal pin profile provides sound material flow among other profiles.

Vinayak et al. [37] investigated the effects of pin profiles on temperature, stir zone and weld power requirements using FEM. According to the results, square pin profile use less power in compare with triangular, cylindrical, rectangular, pentagonal, and hexagonal. Furthermore, frustum pin type produces better results as compared to straight type.

Marzbanrad et al. [9] has considered the influence of pin profiles on mechanical properties and thermal characteristics of AA5083. They investigated the cylindrical and square shapes for the simulation and experiment procedure. A numerical model was developed for investigating the effect of tool pin profiles on material flow, thermal and strain distributions based on thermo-mechanically coupled rigid-viscoplastic 3D FEM. 1120 rpm used for rotational speed, 30 mm/min for traverse speed, and tilting angle of 2° carried out for experiment procedure. Optical microscopy was employed to characterize the microstructures features of the weld and tensile test was performed in order to evaluate the mechanical properties of the joints. From the results they found that square profile produced the higher ultimate strength in compare with cylindrical tool profile and also the grain structure was finer.

Hao and his colleagues [38] investigated the effect of pin profiles on thermal and material flow by numerical modeling. Their simulation is based on computational fluid dynamics to analyze the thermo-physical in FSW. They found that triflat tool produced larger deformation area than conical tool. Moreover, the heat generation of both has not significance differences. However, stir action of triflat is stronger than conical.

Godakh and his colleagues [39] proposed a numerical model for heat generation of different pin profiles. Figure 7 shows the different pin profiles that they used in their work.

They found that square pin profile generated higher temperature, and hexagonal produced lower temperature. Furthermore, their numerical model showed that increasing the rotational speed in constant traversing speed produce higher heat input and vice versa.

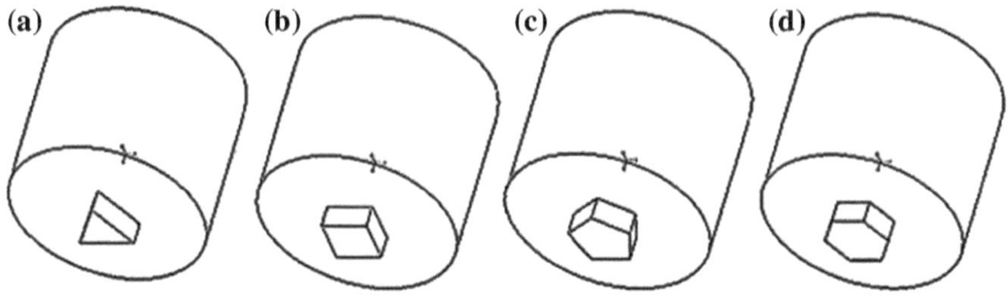

Fig. 7 **a** Triangular pin. **b** Square pin. **c** Pentagonal pin. **d** Hexagonal pin [39]

In the most researches for obtaining the better shapes due to the heat generation and mechanical properties, the finite element analysis is performed. In numerical simulation, three parts of the process are investigated which have their own solving method. Work piece is solved in Eulerian framework, pin is solved in Lagrangian framework, but the stir zone would be solved in Arbitrary Lagrangian-Eulerian (ALE) method [40].

The workpiece is meshed using eight node coupled temperature displacement, brick elements (C3D8RT). The mesh is graded in a way such that there is a higher mesh density around the tool plunge area. This improves the accuracy of the solution around the tool without tremendously increasing the computational time.

The graded meshes are obtained by partitioning the workpiece into smaller cells. Adaptive mesh boundary type lagrangian applied to reduce distortion.

4 Optimization on FSW Pin Profiles

Amongst the experiments, some researchers tried to optimize the FSW parameters. They claimed that it is possible to predict the optimum parameters during FSW process [41–43].

Kalaiselvan and Murugan [41] considered the influence of FSW parameters on tensile strength. They developed a mathematical model to predict the ultimate tensile strength (UTS) of FSW on Al-B_4C composites. The results indicated that the joint fabricated using rotational speed of 1000 rpm, welding speed of 1.3 mm/s, axial force of 10 kN, and the reinforcement of 12% showed higher tensile strength compared with other joints.

Ashok and Murugan [42] investigated rotational speed, welding speed, tool pin profile and percentage of reinforcement which are affected on mechanical properties of friction stir welded Aluminum Matrix Composites (AMC) joints. They developed the regression model to predict the UTS and percent elongation (PE) of FSW on AA6061/AlN_p composite joints. They found that the parameters independently influenced the UTS and PE of FSW joint.

Jagadeesh et al. [43] investigated the tensile test and hardness properties of the twenty-seven joints on Al6061 workpiece of 6.35 mm thick. Moreover, they selected two models (ANOVA, Artificial Neural Network) to predict the tensile strength of friction stir welded AA6061. From the ANOVA results, they concluded that tool rotational speed and axial force are the influential parameters on the mechanical properties. They observed that increasing the parameters such as tool rotational speed, welding speed, and axial force, tensile strength would be increased respectively. Their research also indicates that the weld samples prepared at 2000 rpm, 72 mm/min, 2.5 kN for rotational speed, traveling speed and axial force, respectively with square pin profile, have highest tensile strength around 169.48 MPa.

Most of the previous investigations on optimization of FSW focused on primary parameters such as rotational speed and welding speed with respect to mechanical properties.

Rambabu et al. [44] developed the mathematical model to predict the corrosion resistance of friction stir welded AA2219 aluminum alloy. Their results revealed that the shape of pin has a significant effect on the joint structure and the corrosion properties. Furthermore, the optimized values matches the experimentally determined values.

Optimization of FSW pin profile is accurate with the conforming tests. Regression model and also artificial neural network utilized to predict the tensile strength and evaluate the different parameters. Again from the mechanical and predictive models found that square pin profile produce higher tensile strength in a specified parameters.

5 Conclusion

The most recent research reveals that FSW pin profile have significant effect on material flow and mechanical properties. In addition, the geometry of profiles is effective on heat generation. Some scholars are convinced that there are direct relations between heat generation and hardness, higher temperature, will produce higher hardness. Moreover, some researches claim that other main FSW parameters such as rotational speed have a higher influence on microstructure and mechanical properties. But the role of tool geometry and pin profile is inevitable.

It can also be concluded that among all shapes that are considered, some of them are not compared with each other and they provide defect free welds separately. In the most researches, square pin profile produce sound joints. However, in other studies threaded cylinder or threaded taper provides proper joints. Above all, there is an equal result among all studies in which threaded shapes are most effective on tool performance.

The effect of different pin profiles on mechanical properties is illustrated in Appendix 1.

Acknowledgements The authors would like to acknowledge Universiti Teknologi PETRONAS, Malaysia for the financial support under I-Gen grant (0153DA-135).

Appendix 1: A Summary of Experimental Study on Different Pin Profiles

Material	Plate thickness (mm)	Pin profile	Rotational speed (rpm)	Traverse speed (mm/min)	Tensile strength (MPa)	Hardness (Hv)	Weld joint appearance	References
AA2219	6	Square	1500	45.6	240	88	Good	Elangovan and Balasubramanian [5]
”	”	Square	1600	”	–	105	”	”
”	”	Square	1700	”	–	83	”	”
”	”	Cylindrical	1600	”	100	82	Poor	”
”	”	Conical	1600	”	120	95	No defect	”
”	”	Threaded Cylindrical	1500	”	155	–	Good	”
”	”	Threaded Cylindrical	1600	”	220	–	”	”
”	”	Triangular	1600	”	215	–	”	”
”	”	Triangular	1700	”	150	–	”	”
AA2219	6	Threaded Cylindrical	1600	45.6	220	–	Good	Elangovan and Balasubramanian [6]
”	”	Square	”	22.2	160	96	”	”
”	”	”	”	45.6	240	105	”	”
”	”	”	”	75	180	98	”	”
”	”	Triangular	”	45.6	225	–	”	”
AA6082-O	5	Tri-flute	1200	60	76.67	–	–	Patil and Soman [13]

(continued)

(continued)

Material	Plate thickness (mm)	Pin profile	Rotational speed (rpm)	Traverse speed (mm/min)	Tensile strength (MPa)	Hardness (Hv)	Weld joint appearance	References
”	”	”	”	70	66.33	–	–	”
”	”	”	”	85	56.67	–	–	”
AA6082-O	5	Taper Screw	1200	60	103.33	–	–	”
	”	”	”	70	120	–	–	”
”	”	”	”	75	67	–	–	”
2219-T62 Al alloy	12	Threaded and tapered with three spiral flutes	300–500	60–100	305	105	Fine grain size	Xu et al. [14]
”	”	Threaded and tapered with triangle	”	”	300	93	–	”
AA5083	2.5	Taper Cylindrical	900	16	103	–	–	Ravindra and Surjya [15]
”	”	”	1400	”	122	–	–	”
”	”	”	1800	”	116	–	–	”
”	”	Triangular	900	”	140	–	–	”
”	”	”	1400	”	136	–	–	”
”	”	”	1800	”	123	–	–	”
”	”	Conical	900	”	126	–	–	”
”	”	”	1400	”	138	–	–	”
”	”	”	1800	”	114	–	–	”
”	”	Square	900	”	137	–	Good	”
”	”	”	1400	”	143	–	”	”
”	”	”	1800	”	122	–	–	”

(continued)

(continued)

Material	Plate thickness (mm)	Pin profile	Rotational speed (rpm)	Traverse speed (mm/min)	Tensile strength (MPa)	Hardness (Hv)	Weld joint appearance	References
”	”	Cylindrical	900	”	132	–	–	”
”	”	”	1400	”	147	–	No defect	”
”	”	”	1800	”	140	–	–	”
Az31B Magnesium alloy	3	Cylindrical	630	63	204	70	–	Motalleb-Nejad et al. [17]
”	”	”	800	80	192	68	–	”
”	”	”	1000	100	162	60	–	”
”	”	Conical	630	63	253	80	–	”
”	”	”	800	80	243	77	–	”
”	”	”	1000	100	223	73	–	”
”	”	Threaded Cylindrical	630	63	242	77	–	”
”	”	”	800	80	234	73	–	”
”	”	”	1000	100	212	69	–	”
AA2024-T3	4.8	Triflute	450	180	91	134	–	Trimble et al. [18]
”	”	Square	”	”	90	131	–	”
	”	Cylindrical	”	”	68	123	–	”
AA6061 and AA5086	6	Straight Cylindrical	1100	22	126	–	Defect at the mid thickness region	Ilangovan et al. [20]
”	”	Threaded Cylindrical	”	”	169	–	Defect Free	”

(continued)

(continued)

Material	Plate thickness (mm)	Pin profile	Rotational speed (rpm)	Traverse speed (mm/min)	Tensile strength (MPa)	Hardness (Hv)	Weld joint appearance	References
”	”	Tapered Cylindrical	”	”	163	–	Defect Free	”
AA6082-T6	8	Threaded Cylindrical	710	900	242.1	–	–	Krasnowski et al. [21]
”	”	Triflute	710	900	244	–	–	”
”	”	Cylindrical	710	900	218.4	–	–	”
”	10	Threaded Cylindrical	”	”	233.7	–	–	”
”	”	Triflute	”	”	230.9	–	–	”
”	”	Cylindrical	”	”	215.4	–	–	”
AA6061-T6 and AA2014-T6	–	Straight Cylindrical	900	24	175	–	–	Nadikudi et al. [22]
”	–	Taper Cylindrical	”	”	167	–	–	”
”	–	Stepped Cylindrical	”	”	168	–	–	”
”	–	Square	”	”	183	–	–	”
”	–	Straight hexagon	”	”	172	–	–	”
High-Zinc brass	3	Threaded Cylindrical	450	16	–	64.20	–	Emamikhah et al. [27]
”	”	Threaded taper	”	”	–	50.96	–	”
”	”	Spline	”	”	–	50.84	–	”
	”	Hexahedron	”	”	–	50.49	–	”

References

1. Thomas W et al (1991) International patent no. 1991, PCT/GB92/02203, GB Patent
2. Sulaiman S, Emamian S, Sheikholeslam MN, Mehrpouya M (2013) Review of the effects of friction stir welding speed on stainless steel type 304L. Int J Mater Mech Manuf 1(1):85–87
3. Thompson B (2013) Friction stir welding of 25-mm Thick Steel (cited 2015 23/01/2015). Available from. http://ewi.org/friction-stir-welding-of-25-mm-thick-steel-2/
4. Threadgill PL, Leonard AJ, Shercliff HR, Withers PJ (2009) Friction stir welding of aluminium alloys (cited 2014). Available from http://www.twi-global.com/technical-knowledge/published-papers/friction-stir-welding-of-aluminium-alloys/
5. Elangovan K, Balasubramanian V (2007) Influences of pin profile and rotational speed of the tool on the formation of friction stir processing zone in AA2219 aluminium alloy. Mater Sci Eng A 459(1–2):7–18
6. Elangovana K, Balasubramanian V (2008) Influences of tool pin profile and welding speed on the formation of friction stir processing zone in AA2219 aluminium alloy. Mater Process Technol 200(1–3):163–175
7. Elangovan K, Balasubramanian V (2008) Influences of tool pin profile and tool shoulder diameter on the formation of friction stir processing zone in AA6061 aluminium alloy. Mater Des 29(2):362–373
8. Elangovan K, Balasubramanian V, Valliappan M (2008) Influences of tool pin profile and axial force on the formation of friction stir processing zone in AA6061 aluminium alloy. Int J Adv Manuf Technol 38:285–295
9. Marzbanrad J, Akbari M, Asadi P, Safaee S (2014) Characterization of the influence of tool pin profile on microstructural and mechanical properties of friction stir welding. Metall Mater Trans B 45B:1887–1894
10. Rai R et al (2011) Review: friction stir welding tools. Sci Technol Weld Joining 16(4): 325–342
11. Mishraa RS, Ma ZY (2005) Friction stir welding and processing. Mater Sci Eng R Rep 50 (1):1–78
12. Meilinger Á, Török I (2013) The importance of friction stir welding tool. Prod Process Syst 6 (1):25–34
13. Patil HS, Soman SN (2010) Experimental study on the effect of welding speed and tool pin profiles on AA6082-O aluminium friction stir welded butt joints. Int J Eng Sci Technol 2(5): 268–275
14. Xu W, Liu J, Zhu H, Fu L (2013) Influence of welding parameters and tool pin profile on microstructure and mechanical properties along the thickness in a friction stir welded aluminum alloy. Mater Des 47:599–606
15. Thube RS, Pal SK (2014) Influences of tool pin profile and welding parameters on friction stir weld formation and joint efficiency of AA5083 joints produced by friction stir welding. Int J Innovative Res Adv Eng 1(4):1–8
16. Salari E, Jahazi M, Khodabandeh A, Ghasemi-Nanesa H (2014) Influence of tool geometry and rotational speed on mechanical properties and defect formation in friction stir lap welded 5456 aluminum alloy sheets. Mater Des 58:381–389
17. Motalleb-nejad P, Heidarzadeh TS, Darzi K, Ashjari M (2014) Effect of tool pin profile on microstructure and mechanical properties of friction stir welded AZ31B magnesium alloy. Mater Des 59:221–226
18. Trimble D, O'Donnell GE, Monaghan J (2015) Characterisation of tool shape and rotational speed for increased speed during friction stir welding of AA2024-T3. J Manuf Process 17:141–150
19. Trimble D et al (2015) Friction stir welding of AA2024-T3 plate–the influence of different pin types

20. Ilangovan M, Rajendra Boopathy S, Balasubramanian V (2015) Effect of tool pin profile on microstructure and tensile properties of friction stir welded dissimilar AA 6061–AA 5086 aluminium alloy joints. Defence Technol 11(2):174–184
21. Krasnowski K, Hamilton C, Dymek S (2015) Influence of the tool shape and weld configuration on microstructure and mechanical properties of the Al 6082 alloy FSW joints. Arch Civ Mech Eng 15(1):133–141
22. Nadikudi BKB et al (2015) Formability analysis of dissimilar tailor welded blanks welded with different tool pin profiles. Trans Nonferrous Met Soc China 25(6):1787–1793
23. Sato YS, Urata M, Kokawa H (2002) Parameters controlling microstructure and hardness during friction-stir welding of precipitation-hardenable aluminum alloy 6063. Metall Mater Trans A 33A:625–635
24. Muhsin JJ, Tolephih MH, Muhammed AM (2012) Effect of friction stir welding parameters (rotation and transverse) speed on the transient temperature distribution in friction stir welding of AA 7020-T53. ARPN J Eng Appl Sci 7(4):436–446
25. Mironov S, Inagaki K, Sato YS, Kokawa H (2015) effect of welding temperature on microstructure of friction-stir welded aluminum alloy 1050. Metall Mater Trans A 46A: 783–790
26. Manvatkar V et al (2015) Cooling rates and peak temperatures during friction stir welding of a high-carbon steel. Scripta Mater 94:36–39
27. Emamikhah A, Abbasi A, Atefat A (2014) Effect of tool pin profile on friction stir butt welding of high-zinc brass (CuZn40). Int J Adv Manuf Technol 71:81–90
28. Mohanty HK, Mahapatra MM, Kumar P, Biswas P, Mandal NR (2012) Effect of tool shoulder and pin probe profiles on friction stirred aluminum welds—a comparative study. J Mar Sci Appl 11(2):200–207
29. Reza-E-Rabby M, Reynolds AP (2014) Effect of tool pin thread forms on friction stir weldability of different aluminum alloys. Procedia Eng 90:637–642
30. Venkata Rao C, Madhusudhan Reddy G, Srinivasa Rao K (2015) Microstructure and pitting corrosion resistance of AA2219 Al–Cu alloy friction stir welds—effect of tool profile. Defence Technol 11:123–131
31. Venkata Rao C, Madhusudhan Reddy G, Srinivasa Rao K (2015) Influence of tool pin profile on microstructure and corrosion behaviour of AA2219 Al-Cu alloy friction stir weld nuggets. Defence Technol 11:123–131
32. Awang M, Mucino VH, Feng Z, David SA (2005) Thermo-mechanical modeling of friction stir spot welding (FSSW) process: use of an explicit adaptive meshing scheme. SAE Technical paper, 2006, 2005(1), p 1251
33. Song M, Kovacevic R (2003) Thermal modeling of friction stir welding in a moving coordinate system and its validation. Int J Mach Tools Manuf 43:605–615
34. Kheireddine AH, Khalil AA, Ammouri AH, Kridli GT, Hamade RF (2013) An experimentally validated thermo-mechanical finite element model for friction stir welding in carbon steels. World Acad Sci Eng Technol 7(4):309–312
35. Buffa G, Hua J, Shivpuri R, Fratini L (2006) Design of the friction stir welding tool using the continuum based FEM model. Mater Sci Eng A 419:381–388
36. Patel JB, Patil HS (2014) Simulation of peak temperature & flow stress during FSW of aluminium alloy AA6061 for various tool pin profiles. Int J Mater Sci Eng 2(1):67–71
37. Malika V, Sanjeev NK, Suresh Hebbarb H, Kailasa SV (2014) Investigations on the effect of various tool pin profiles in friction stir welding using finite element simulations. Procedia Eng 97:1060–1068
38. Su H et al (2015) Numerical modeling for the effect of pin profiles on thermal and material flow characteristics in friction stir welding. Mater Des 77:114–125
39. Gadakh VS, Kumar A, Patil GJ (2015) Analytical modeling of the friction stir welding process using different pin profiles. Weld J 94(4):115S–124S
40. Chiumenti M, Cervera M, de Saracibar CA, Dialami N (2013) Numerical modeling of friction stir welding processes. Comput Methods Appl Mech Eng 254:353–369

41. Kalaiselvan K, Murugan N (2013) Role of friction stir welding parameters on tensile strength of AA6061−B4C composite joints. Trans Nonferrous Met Soc China 23:616–624
42. Ashok Kumar B, Murugan N (2014) Optimization of friction stir welding process parameters to maximize tensile strength of stir cast AA6061-T6/AlNp composite. Mater Des 57:383–393
43. Jagadeesh Chandra Prasad P, Hema P, Ravindranath K (2014) Optimization of process parameters for friction stir welding of aluminum alloy AA6061 using square pin profile. Int J Mech Eng Robot Res 3(2)
44. Rambabu G et al (2015) Optimization of friction stir welding parameters for improved corrosion resistance of AA2219 aluminum alloy joints. Defence Technol 11:330–337

Friction Stir Welding of Polymers: An Overview

Adeel Zafar, M. Awang and Sajjad Raza Khan

Abstract Friction stir welding, a solid state joining process is an already proven technique for the welding of metals. Lately, it has been employed on polymers, which also have shown efficacious joining results. Due to recent consideration of this process on polymers, a limited body of literature on friction stir welding of polymers exists. This paper presents an overview of the state-of-the-art of the process on similar and dissimilar polymers. The process parameters which are significant for weld strength and various approaches to maximize it, such as external heat induction, tool modification, and material flow are central to this discussion. Finally, the conclusions drawn with future recommendations are reflected as well.

1 Introduction

Considering the light weight property of polymeric materials, their use has grown widely in various industries, mainly in electronics, automotive, aerospace, and packaging [1]. Due to the dissimilarity in materials and complexity of joining parts, the increased use of any material at the same time increases the importance of the joining process. Several joining techniques for similar and dissimilar polymeric materials are employed in many industries. The invention of modern welding processes such as friction welding, resistance, ultrasonic, laser, induction, and microwave welding has widely replaced conventional joining processes (adhesive and mechanical fastening) [2]. However, these modern processes require high initial investment which limits their scope of application. Better mechanical properties

A. Zafar · M. Awang (✉) · S.R. Khan
Department of Mechanical Engineering, Universiti Teknologi PETRONAS, 32610 Seri Iskandar, Perak, Malaysia
e-mail: mokhtar_awang@utp.edu.my

A. Zafar
e-mail: adeel_ze@ymail.com

S.R. Khan
e-mail: sajjadrkhan@yahoo.com

M. Awang (ed.), *2nd International Conference on Mechanical, Manufacturing and Process Plant Engineering*, Lecture Notes in Mechanical Engineering,
DOI 10.1007/978-981-10-4232-4_2

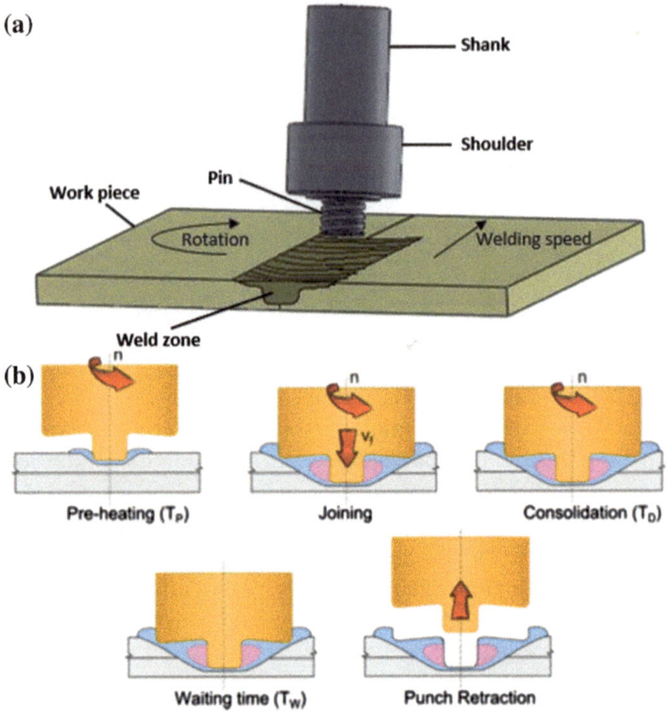

Fig. 1 A schematic representation of **a** FSW and **b** FSSpW process [1]

with low investment cost have been obtained by friction stir welding (FSW) and friction stir spot welding (FSSpW). FSSpW is the variant of FSW by the elimination of transverse part and involves plunging and retraction at one spot. A schematic view of both processes is shown in Fig. 1. These friction based processes have many other advantages such as solid state welding, flexibility, and insignificant heat-affected zone (HAZ) in polymers [3].

Among three kinds of polymeric materials; thermoplastics, thermosets, and elastomers, only thermoplastics are the weldable polymers. It is due to their ability to be reshaped after heating below their degradation temperature. Examples of such polymeric materials include Polyvinyl chloride (PVC), Polystyrene (PS), Acrylonitrile Butadiene Styrene (ABS), Polymethyl methacrylate (PMMA), low-density and high-density polyethylene (PE). Others are Polypropylene (PP), Poly tetra fluoro ethylene (PTFE), nylon-6 (PA 6), and polycarbonate (PC) [4]. Due to different rheological properties of theses polymers particularly melt viscosity, their flow behavior is also different. This difference leads to the diversity of process parameters as well as pin profile in FSW process.

The joining mechanism in welding of polymers is based on molecular diffusion and chain entanglement which starts at the molten stage of the polymer. After cooling and solidification of the weld, residual stresses due to squeeze flow will remain in the weld [5]. Grewell [5] has mentioned that a high squeeze flow rate will lead to a certain molecular alignment direction which ultimately influences the weld quality and may result in a weak weld. Moreover, the degree of welding also depends on the kind of polymer used, its properties, and the joining parameters.

The present review focuses on using the FSW and FSSpW to join similar and dissimilar polymers. Firstly, it describes the problems and challenges surmounted to friction stir weld the polymeric materials. This is followed by a detailed description of progress so far achieved on the FSW of polymers. Finally, conclusions are made with a particular view on future research challenges and directions.

2 Friction Stir Weldability of Polymers

FSW of polymers is critical due to different physical and rheological differences present in each polymer. The achievement of optimum process parameters becomes more challenging when it comes to weld polymers which have low melt viscosity such as nylon-6. Rotational speed is the major process parameter in FSW process which varies for each kind of material depending on their physical properties [6]. A higher rotational speed results in the degradation of the polymer, whereas lower rotational rate gives poor mixing thus producing voids in stir zone. Therefore, the need to investigate the optimum parameters for each polymer is vital.

On the other hand, squeeze-out of plasticized or semi-molten material is also an important consideration to achieve maximum weld strength as it may result in flash formation and eventually a poor surface finish [7]. An example of over plasticized material's squeeze out as observed in PC is shown in Fig. 2. In addition to that, an excessive flow-out of the material may result in tunnel defect which will ultimately reduce the weld quality. The uniform mixing of material during the stirring process has a direct relationship with weld quality [8]. On the other hand, uniformity in material flow or mixing depends on the geometry of pin profile [9]. Therefore, optimum pin profile for uniform mixing should also be evaluated to widen the scope of FSW on polymers.

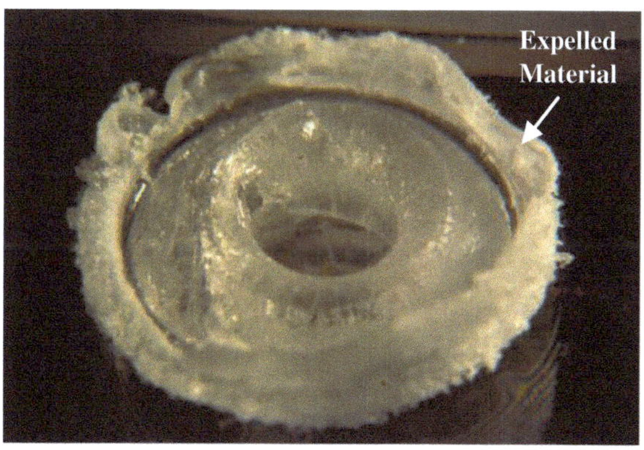

Fig. 2 Expulsion of plasticized material during FSW of PC [10]

3 The Research Progress in Friction Stir Welding of Polymers

FSW and FSSpW of polymer have continued to develop rapidly in recent years with advances made in tool design, external heat introduction, and different submerging conditions. Moreover, material flow during welding has also been studied using marker insert technique. This review presents a comprehensive review of all investigations made to maximize the weld strength. Table 1 is the sum of the findings on FSW and FSSpW which was performed on different polymers.

3.1 *Optimum Welding Parameters for the FSW of Polymers*

The FSW require very few process parameters (rotational speed, welding speed, tool dimension, and plunge depth) [24] and each parameter has vital roles in heat generation and stirring. The optimization of parameters is significant to produce an acceptable joint strength. Due to different rheological and physical differences among polymers, they were welded at diverse FSW parameters. Besides, it was discovered that workpiece thickness profoundly influences the welding parameters. Also, researchers have reported different joining parameters for various levels of thickness of the same material. For instance, Azarsa et al. [21] welded 10 mm thick high-density polyethylene (HDPE) at 1400 RPM rotational speed and 25 mm min^{-1} welding speed with the joining strength almost equal to the base material. Bozkurt [6] welded thin (4 mm) HDPE sheet at higher rotational and welding speed which is 3000 RPM and 115 mm min^{-1} respectively. Although there is a significant difference in joining parameters of both studies, the joint strength in thin sheets is only 10% lower when compared to thick sheets. The decrease in strength could be due to higher rotational rate which generated high temperature i.e., 165 °C which is higher than HDPE melting point (132 °C).

The similar difference of process parameters (rotational and welding speed) has been found during the FSW of nylon-6; however, the difference of joint strength is negligible at different thicknesses of the workpiece. Inaniwa et al. [11] joined the 5 mm thick nylon-6 at 440 RPM rotational speed and 40 mm min^{-1} welding speed with a tool shoulder-workpiece gap of 0.1 mm. Panneerselvam et al. [7] welded 10 mm thick nylon-6 sheet at 1000 RPM rotational speed and 10 mm min^{-1} welding speed with a tool shoulder-workpiec gap of 0.5 mm. In a recent study, Zafar et al. [25] further increased the nylon-6 sheet thickness to 16 mm and successfully welded without any gap (zero plunge depth) between tool shoulder and workpiece. However, in the study by Zafar et al. [25] the rotational speed and welding speed was reduced to lower levels, i.e. 300 RPM and 25 mm min^{-1} respectively. Overall, the joint strength at any thickness remained between 30 and 40% of the base material. The difference in the welding speed in all three studies on nylon-6 is minimal. The rotational speed difference may be attributed to the tool

Table 1 States of the art in polymer friction stir welding

Author (s)	Base material	Plate thickness (mm)	Tool/pin profile	UTS (MPa)	Joint efficiency (%)
Inaniwa et al. [11]	Nylon-6	5	Threaded	67.1	35
	High-density polyethylene (HDPE)	5	Threaded	31.9	70
	Polyvinyl Chloride (PVC)	5	Threaded	66.5	45
Panneerselvam et al. [7]	Nylon-6	10	Threaded	34.8	40
Zafar et al. [12]	Nylon-6	16	Threaded	27.22	32
Sadeghian et al. [13]	Acrylonitrile butadiene styrene (ABS)	8	Cylindrical	41.42	99.1
			Conical	39.30	94
Bagheri et al. [14]	ABS	5	Threaded	32.62	88.76
Pirizadeh et al. [15]	ABS	5	Double shoulder tool with convex pin	20.7	60.6
			Double shoulder tool with simple pin	15.58	45.6
Mendes et al. [16]	ABS	6	Stationary shoulder with conical threaded pin	29.48	67
Oliveira et al. [17]	Polymethyl methacrylate (PMMA)	3	Refill friction stir spot welding tool	9.5	11.87
Ahmadi et al. [18]	PP	4	Simple cylindrical conical	5.7	–
			Threaded conical	3.84	–
Kiss et al. [19]	PP	15	Traditional milling tool	11.5	50
Bozkurt [6]	HDPE	4	–	19.4	86.2
Gao et al. [20]	HDPE	4	Threaded cylindrical	12.3	–
Azarsa et al. [21]	HDPE	10	Threaded	33.76	95.69
Arici et al. [22]	Medium-density polyethylene (MDPE)	5	Cylindrical	20	100
Aydin [23]	Ultra-high molecular weight polyethylene (UHMW-PE)	4	Threaded	28.5	89

shoulder-workpiece gap because a higher difference requires a higher rotational speed to generate heat. It was observed that when the gap between tool shoulder and workpiece is helpful to generate low heat at the workpiece surface, it also promoted flash formation. This is the reason that Zafar et al. [25] concluded in their study that the insignificant flash formation at the workpiece surface as observed in Fig. 3, represents the cross-sections of best-welded joints.

Another polymer of nylon family, nylon-66, has been friction stir spot welded (FSSpW) by Husain et al. [26] and it showed an increase in the welding strength with the increase of rotational speed at a limit of 1570 RPM. The further increase in rotational speed (higher than 1570 RPM) resulted in the decrease of strength due to higher heat generation. However, a fall in joint strength was observed with the increase of welding speed. The maximum relative joint strength was pegged at 55% in comparison with the base material.

While achieving the operative conditions in FSSpW of thermoplastic materials, various approaches have been developed to determine the parameters relation with joint strength. For example, Paoletti et al. [1, 10, 27] developed a prototypal setup to observe the effect of plunging force, torque, and tool temperature during the FSSpW of 3 mm thick Polycarbonate (PC). The results were used to develop an Artificial Neural Network to predict the maximum values of parameters as well as the shear strength of the joint. According to experimental results, the shear strength of welded joint is primarily influenced by the plunge rate, the rotational speed, and the waiting time (post-weld cooling time). The maximum lap shear strength was obtained at lower values of tool rotational speed, plunge rate and at high values of pre-heating, as well as dwell and waiting times.

As far as dissimilar FSW of polymers is a concern, it is not well investigated compared to metals [28]. In a reported work, Dashatan et al. [29] showed the feasibility of dissimilar polymer joining by FSSpW of PMMA and ABS. In their study, it was found that welding parameters, particularly rotational speed and plunge rate, have a significant effect on the joint strength. Lap shear strength results indicated that 800 RPM is an optimum rotational speed as at below and above this value; a decrease in strength was observed. Low rotational speed caused lack of heat generation, while high rotational speed led to extra heat, which squeezed-out the over-plasticized material. Tool plunge rate had an inverse relation with the

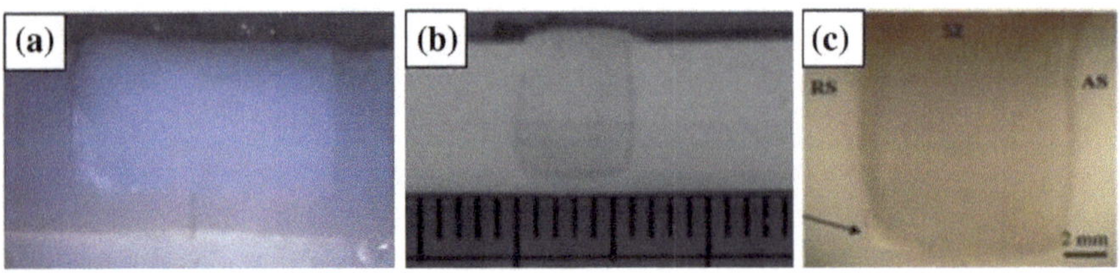

Fig. 3 Cross sections of friction stir welded nylon-6 specimens **a** 5 mm thick [11] **b** 10 mm thick [7] **c** 16 mm thick [25]

strength of the weld because an increase in the tool plunge rate decreased the weld strength.

A similar role of welding parameters was reported by Eslami et al. [30] in their attempt to join the PP with PE. They friction stir welded the workpieces using a novel stationary shoulder tool. Results showed maximum tensile strength at 2500 RPM rotational speed and 100 mm min^{-1} welding speed. However, the samples welded under same welding parameters did not show a good reproducibility as had been observed previously in metal welding. This feasibility study of PP-PE, and PMMA-ABS joining by Dashatan et al. [29] however indicates the practicability of dissimilar joining in polymers. Further investigations on other dissimilar polymers joining should be conducted due to their wide engineering applications.

The plunge depth in FSW of polymer is crucial because a higher depth can exaggerate the heat at workpiece surface due to the thermal insulator nature of polymer. Bilici and Yukler [31] have mentioned the plunge depth effect in their study on HDPE using FSSpW. As depicted in Fig. 4, the joint strength increases to a certain level of plunge depth. The further increase in plunge depth resulted in the squeeze out of polymer which ultimately lowered its strength value. Furthermore, they observed two fracture modes in lap shear tests; cross nugget failure and pull nugget failure. The welded joint with higher strength fractured with pull nugget morphology was observed as well.

In conclusion, among all process parameters, rotational speed had a significant effect on joint strength. Bozkurt [6] mentioned 73.85% contribution of rotational speed to the overall welding parameters. As can be seen in Fig. 5, the least contribution was made by a tilt angle. However, no general trend can depict the relationship between welding and rotational speed. Figure 6 represents an optimum rotational speed and welding speed for different polymers recommended by researchers. Overall, it is clear that polyethylene is welded at a higher rotational and welding speed. Whereas nylon-6, due to its low melt viscosity, requires a lower rotational and welding speed.

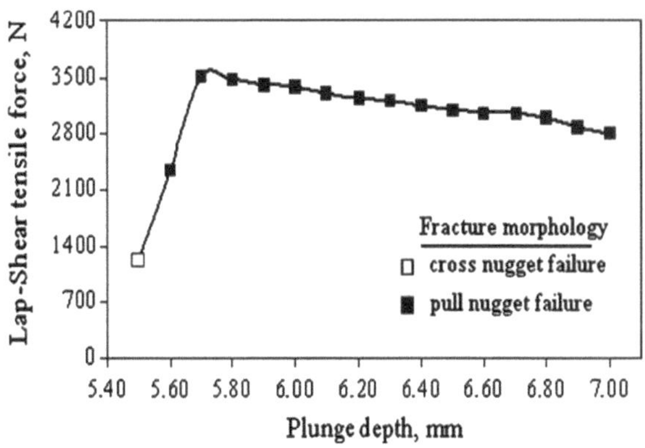

Fig. 4 The influence of tool plunge depth on lap-shear tensile strength [31]

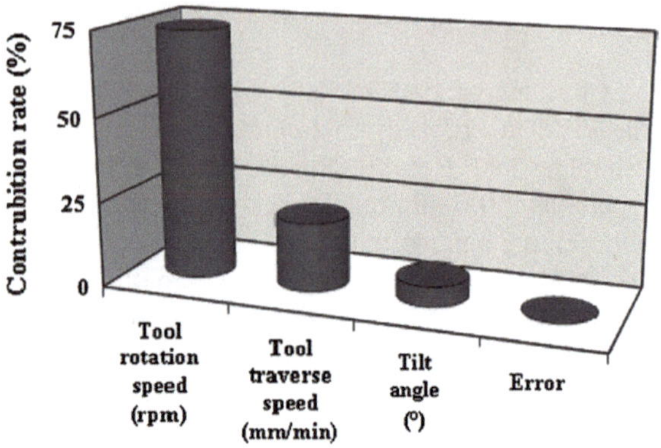

Fig. 5 Welding parameters' contribution rate [6]

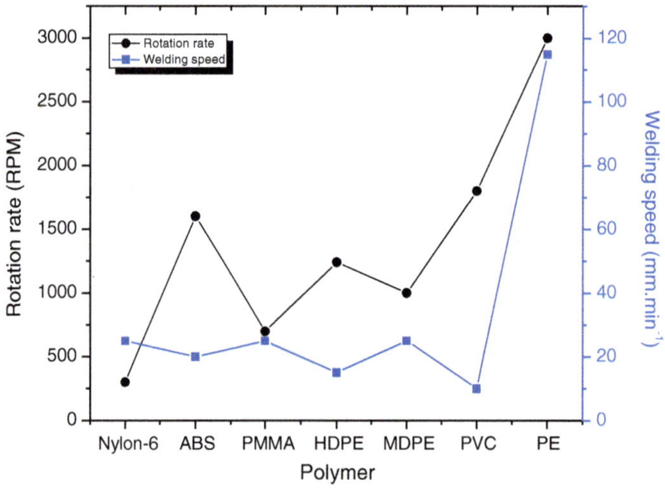

Fig. 6 Summary of optimum joining parameters for different polymers

3.2 *Role of Tool Design on FSW of Polymers*

In FSW, pin and shoulder geometry of the tool profoundly affect the weld quality [32]. Considering the importance of tool geometry, wide research on tool design has also been performed. Bilici et al. [33] studied the effect of six different commonly used pins on PP sheets in the FSSpW process. The types of pins studied included; straight cylindrical, tapered cylindrical, threaded cylindrical, square, triangular, and hexagonal pins as shown in Fig. 7. Results showed that the threaded pin with 0.8 mm pitch length has the highest tensile strength. Moreover, it was also observed

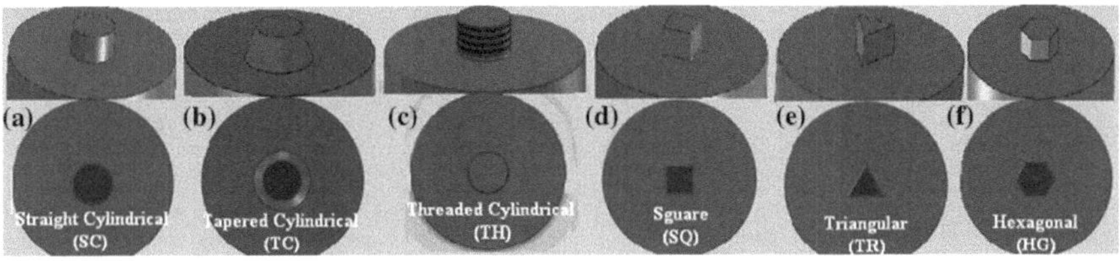

Fig. 7 Different pin profiles of tool for FSW [33]

that the pitch length of a pin type has a reasonable effect on the welding strength and it decreased with the decrease of pitch length.

Similarly, Ahmadi et al. [18] also studied the effect of pin profile on carbon fiber reinforced polypropylene composite in lap-joint configuration. They used four different pin profiles; threaded cylindrical, threaded tapered, straight cylindrical and straight tapered. Experiments conducted at fixed optimum parameters showed best tensile results for tapered threaded pin among the four pin profiles. They believed that it was due to the wider contact surface of the tapered pin with the specimen which resulted in high friction. On the other hand, the threads on pins caused a large amount of turbulence on the weld seam, which resulted in the better mixing of molten material and consequently, a better quality of the weld.

Besides pin profile, the effect of shoulder profile has been investigated on ABS by Sadeghian et al. [13]. They used a specially designed concave shoulder with straight cylindrical and conical pins. This particular design of shoulder was made to restrict the generated heat of friction within the weld area. A stainless steel blade was also placed on the upper surface of the ABS sheets to prevent the top of the weld from undercutting thus producing a good surface finish. The tensile strength of specimens showed effective results as the highest weld efficiency was pegged at 101 and 99% in the conical and the cylindrical pinned tools, respectively.

The novel designs of tools have also been employed. In a preliminary study, Oliveira et al. [17] joined thin PMMA sheets using refill friction stir spot welding. This process is efficient in a way that a key-hole will be filled by a non-consumable FSW tool. The schematic of the tool is shown in Fig. 8a. The details of the tool design and its working principle have been described in the study conducted by Goushegir et al. [34]. The maximum achieved joint lap-shear strength was about 9.5 MPa. Comparing with other joining techniques including microwave welding, thermal bonding, and ultrasonic welding, they found that friction spot welding gave a higher strength in a short joining time. The graphical representation is given in Fig. 8b. In another study, a novel robotic platform was developed by Mendes et al. [35] to increase the flexibility of the process. The platform consisted of three major groups of hardware; a robotic manipulator, an FSW tool, and support for the FSW tool. The setup as a preliminary studied on ABS showed comparable joint strength.

Having reviewed the literature, it is apparent that the pin profile plays a decisive role in determining the strength of the joint. The threaded pin, due to its ability to

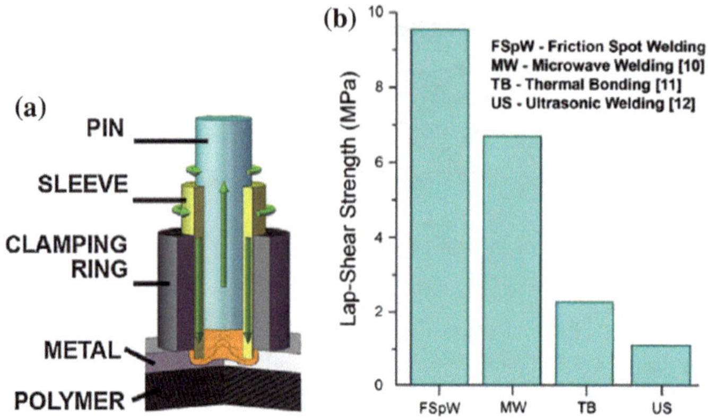

Fig. 8 **a** Schematic illustration of the refill friction stir spot welding tool. **b** Lap-shear strength comparison of FSSpW with other joining techniques [34]

adequately mix the plasticized material has shown good welding results. Moreover, a high surface area of threaded pin generates a higher frictional heat which is an essential pre-condition to produce a weld.

3.3 Effects of External Heat Induction

Due to thermal insulator nature of polymers, it is always difficult to conduct the shoulder generated heat to the base of workpiece. It may cause a heterogeneous mixing near the base, especially in thick sheets. This challenge was addressed by adopting different approaches. For instance, Aydin [23] pre-heated the ultra-high molecular weight polyethylene (UHMW-PE) sheets before employing the FSW. Pre-heating at 50 and 80 °C from the bottom of sheet was performed in the furnace. Comparing the tensile values of pre-heated specimens with non-heated specimens, the research found that the specimens heated at 50 °C temperature achieved a 17% higher strength. On the contrary, the strength of non-heated specimen was 72% of the parent material's strength. Although pre-heating showed good joining results, an additional heating step not only affects the simplicity of the process but also increases the process time.

The in situ heating during the FSW process could be a good approach to maintaining the process short joining time. Bagheri et al. [14] employed a heated shoe called 'Hot shoe' to improve the welding strength as well as to eliminate the defects in ABS joints. Hot shoe made from steel is an externally heated stationary shoulder which moves linearly along the tool. This hot shoe not only heats the specimen but also keep the plasticized material within the weld zone. Efficient results were obtained at higher rotational speeds, high shoe temperature, and lower welding speed. Lower welding speeds gave sufficient time for mixing and hence good welding strength was achieved. The highest tensile strength was almost 89% of the base material strength. In another study, Mostafapour and Asad [36] also used hot shoe to weld nylon-6. Results showed good welding results as tensile strength was almost equal to that of the base material.

Another approach to generating in situ pre-heat is given by Vijendra and Sharma [37]. However, in this technique, only the tool is heated through induction heating process, and constant temperature is maintained throughout the process. For this purpose, an induction coil encircling the tool was used and supplied with alternating current. This investigation showed remarkable results in which joint strength near to the strength of the base material was achieved at a tool-pin temperature of 45 °C, and a rotational speed of 2000 RPM. This approach of tool induction heating has some advantages over other heating methods. For instance, the shoe heating technique may not be used for curvilinear welds for the reason that pre-heating of tool might not perform longer welds because of dissipation in tool temperature, and external heating of tool will probably lead to erratic heating of the base material.

3.4 Submerged FSW of Polymers

Various researches into the FSW of metals in submerged condition have been conducted, and the results have been promising. [38]. Similar research is also required on polymers. However, in literature, a limited body of knowledge exists. Gao et al. [20], employed FSW on HDPE sheets under water and air and compared their tensile strengths. Experiments were performed at optimum welding parameters using a threaded cylindrical pin. Tensile test results showed that the welding performed under water had a higher tensile value when compared with welding performed in air. The maximum strength value under water was 12.3 Mpa, whereas it reduced to 9.6 Mpa in air. All the fractures were observed in the heat affected zone (HAZ). They believed that in a submerged condition, the water not only controls the peak temperature but also affect the thermal cycle of the weak part, resulting in the increase of joint strength.

3.5 Microstructure Evolution

Welding parameters have a direct relationship with the microstructure of the material and hence the quality of the weld. During FSW, the workpiece undergoes intense heating which results in microstructural changes and different zones formation. In metals FSW, four different zones: the weld nugget (WN), thermo-mechanically affected zone (TMAZ), heat-affected zone (HAZ) and base material (BM) are observed [3]. However, in the case of polymers, HAZ is not always present. Moreover, the microstructure observed in the welded zone is different from metals. Kiss et al. [39] studied the morphology of weld seams of friction stir welded

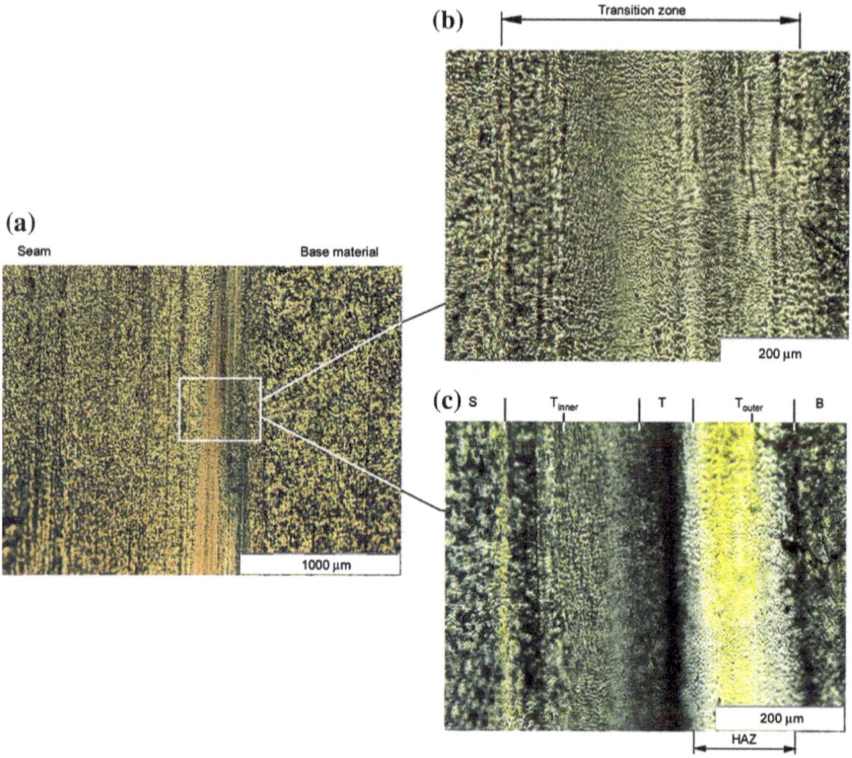

Fig. 9 **a** Spherulitic structure in the seam and base material. **b** and **c** Supermolecular structure in the border region [39]

PP sheets. As shown in Fig. 9, the results of optical microscope showed spherulitic structure at the center and base material. The structure at the center was similar to that of the base material. They believed that to achieve good weld quality; the welds should have less complexity in morphology with small seam width.

3.6 *Elimination of Root Defect in FSW of Polymers*

After the successful application of FSW on polymers, attempts have been made to eliminate the associated defects to improve the welding strength. Root defect which occurs due to incomplete penetration of tool has a direct influence on the tensile strength of weld [40]. In most tensile failures, it has been found that fractures were initiated by root defects [41]. Various approaches such as double shoulder scheme and double passes of the tool have been used to eliminate this defect. Arici et al. [22] employed the double passes of the tool on medium density polyethylene (MDPE) sheets. In their process, welding was performed in two steps. In the first

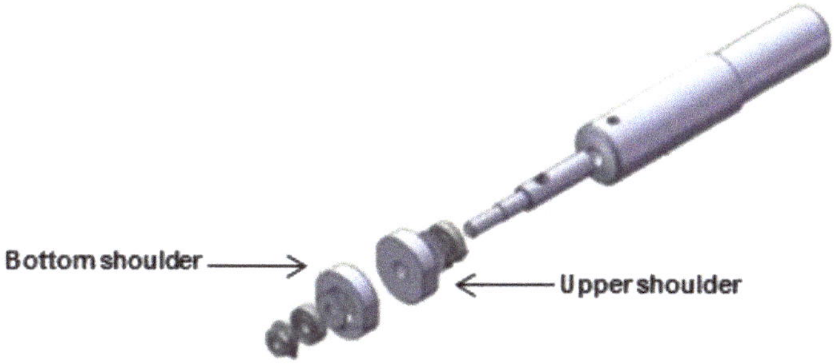

Fig. 10 A novel design double shoulder tool [15]

step, half thickness of specimen was friction stir welded, while the remaining half was welded in the second step by turning back the specimen in the fixture. Results showed the root defect free specimen with an increased tensile value.

In another study, Pirizadeh et al. [15] used a newly designed tool which consisted of two shoulders and was capable of contacting upper and lower surface of the workpiece during the process. The tool is shown in Fig. 10. This tool was designed to eliminate the root defect in samples. As a working principle, shoulders were kept stationary during the process by placing the ball bearings and thus there was only pin rotation. The experiments were conducted on 5 mm thick ABS sheets. After welding, no root defect or back slit was observed. Tensile results showed that the highest tensile value was almost 60% of the base material.

3.7 *Effects of Crystallinity on the Properties of Polymer*

The crystallinity of polymer indicates the extent of structural order. A fully crystalline material has atoms or molecules in an ordered and periodic manner [42]. The degree of crystallinity has a significant influence on the properties of materials. It includes tensile strength, stiffness, yield point, hardness, and impact strength. Increase in crystallinity has shown an increase in tensile strength and hardness with the loss of impact strength [43–45]. The degree of crystallinity is usually measured regarding percent crystallinity by differential scanning calorimetry. Percent crystallinity is an overall crystalline content in comparison to its amorphous component [46]. On the other hand, differential scanning calorimetry (DSC) is a method which measures exothermic or endothermic heat flow of a material as a function of time or temperature. Polymer crystallinity can be found with DSC by measuring the heat of polymer fusion. This heat gives the percent crystallinity by rationing against the heat of fusion for a 100% or fully crystalline specimen of the same material [47].

Gao et al. [20] measured percent crystallinity of friction stir welded HDPE samples at different conditions such as base material (BM), HAZ, thermo-mechanical affected zone (TMAZ), and weld nugget (WN). These regions are

Fig. 11 Different regions formed during FSW [20]

shown in Fig. 11. DSC curves were used to determine percent crystallinity using the formula given below;

$$\mathbf{Wc} = \frac{\mathbf{\Delta Hm}}{\mathbf{\Delta H^{\circ}m}} \times 100\%$$

Results showed a decrease in crystallinity while approaching from BM to WN. They believed that FSW was a fast welding process which resulted in a decrease of crystallinity due to rapid cooling of the weld zone. It is important to mention here that rapid cooling causes the loss of crystallinity in the material [48].

Another study was performed on friction stir welded PP specimens by Kiss et al. [19] using a Perkin-Elmer differential scanning calorimeter at a heating rate of 10 K min^{-1}. The results showed almost the same trend as reported by Gao et al. [20]. Weld Nugget (WN) was found to have the lowest crystallinity as compared to other regions. They believed that it was due to a larger tool size compared to the WN which rapidly absorbed the heat from WN. As a result, fast cooling caused the lower crystallinity in WN. The region at a distance from the tool had more time for cooling which made it a comparatively high crystalline.

3.8 Material Flow During FSW Process

The understanding of the material flow is important in preventing blowholes and other defects which may form during the stirring process [49]. These defects also affect post weld microstructure and mechanical properties. Therefore, the study of material flow in both sides; advancing side (AS) and retreating side (RS) of the tool is critical. Limited literature exists on the material flow during FSW of polymers. Simoes et al. [50] studied material flow on PMMA sheets using two different tools. Tool 1 had a small diameter of shoulder with a conical profile, whereas tool 2 had a large diameter with a conical profile near the center. Comparing their results with Arbegast [51] flow model study on metal, they believed that the material flow in polymers is different from that of metals. Also, they reported that the interface of pin plunged zone and the base material remained straight which showed no material

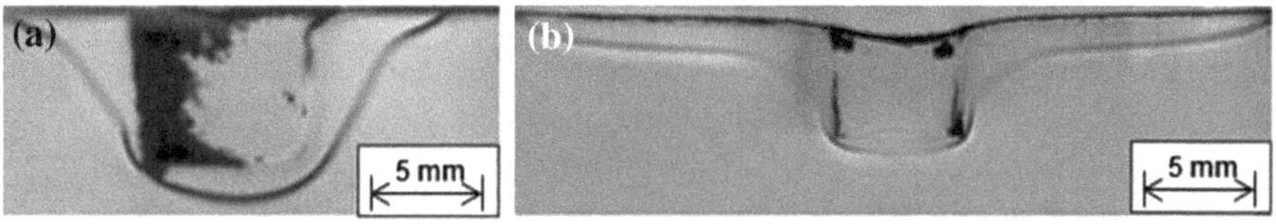

Fig. 12 Backlight images of Friction stir welded PMMA (transparent in nature) **a** Tool 1 **b** Tool 2 [50]

flow from stir zone to the base material. The images of both specimens are shown in Fig. 12.

A similar observation of material flow was noted during FSW of nylon-6 by Zafar et al. [52]. Also, during visual analysis made at different sections of welded specimens, an overall uniform material mixing was observed. The vertical, as well as the horizontal displacement in which material moved from AS to RS and vice versa, show the successful feasibility of the process on polymers.

4 Conclusions

Friction stir welding, after its successful application on metals, has been successfully utilized for polymers. A variety of thermoplastic polymers has been friction stir welded and investigated regarding their micro-mechanical and thermal properties. We found that depending on their physical and rheological properties, the welding parameters differ from polymer to polymer. Polymers with high melting temperature and viscosity, require higher rotational speed and low welding speed to achieve sufficient heat and eventually good weld strength. Moreover, to enhance the welding strength, hot-shoe utilization or an in situ induction heating of the tool has shown good results. In the evaluation of appropriate pin profile, we saw that it varies from polymer to polymer. For example, the conical pin is reported as best profile for acrylonitrile butadiene styrene and high-density polyethylene, whereas the threaded cylindrical pin profile is considered most suitable for polypropylene friction stir welding. Furthermore, we found that different pin profiles also led to different material flow. However, a uniform material flow which is a bit like flow in metals, have also been observed.

5 Recommendations

The joining of polymers using FSW is still under development and therefore has not yet been fully employed in the industry. The results showed that it is a potential welding process; however, some further modifications and investigations are required to make it fully utilized. There should be a comprehensive study on the plunge depth relation with joint strength. More work is needed to understand the welding parameters-microstructure relation. Also, there is a need to understand the bonding mechanism of dissimilar polymers so that the scope of the process could be extended to other polymers. Furthermore, the application of the process for different joint specifications while maintaining the process simplicity and flexibility could be a significant achievement to implement the process in the industry.

Acknowledgements The authors would like to thank Universiti Teknologi PETRONAS for providing a research platform with financial assistance.

References

1. Lambiase F, Paoletti A, Di Ilio A (2015) Mechanical behaviour of friction stir spot welds of polycarbonate sheets. Int J Adv Manuf Technol 80(1–4):301–314
2. Ageorges C, Ye L, Hou M (2001) Advances in fusion bonding techniques for joining thermoplastic matrix composites: a review. Compos A Appl Sci Manuf 32(6):839–857
3. Mishra RS, Mahoney MW (2007) Friction stir welding and processing. ASM International
4. Whelan A (2012) Polymer technology dictionary. Springer Science & Business Media
5. Grewell D, Benatar A (2007) Welding of plastics: fundamentals and new developments. Int Polym Proc 22(1):43–60
6. Bozkurt Y (2012) The optimization of friction stir welding process parameters to achieve maximum tensile strength in polyethylene sheets. Mater Des 35:440–445
7. Panneerselvam K, Lenin K (2014) Joining of Nylon 6 plate by friction stir welding process using threaded pin profile. Mater Des 53:302–307
8. Lorrain O et al (2010) Understanding the material flow path of friction stir welding process using unthreaded tools. J Mater Process Technol 210(4):603–609
9. Morisada Y et al (2011) Three-dimensional visualization of material flow during friction stir welding by two pairs of X-ray transmission systems. Scripta Mater 65(12):1085–1088
10. Paoletti A, Lambiase F, Di Ilio A (2016) Analysis of forces and temperatures in friction spot stir welding of thermoplastic polymers. Int J Adv Manuf Technol 83(5–8):1395–1407
11. Inaniwa S et al (2014) Application of friction stir welding for several plastic materials. In: Proceedings of the 1st international joint symposium on joining and welding, Osaka, Japan, 6–8 November 2013. Woodhead Publishing
12. Zafar A et al (2015) Effect of double shoulder tool rotational speed on thermo-physical characteristics of friction stir welded 16 mm thick nylon6. Appl Mech Mater J 799–800: 251–255
13. Sadeghian N, Givi MKB (2015) Experimental optimization of the mechanical properties of friction stir welded Acrylonitrile Butadiene Styrene sheets. Mater Des 67:145–153
14. Bagheri A, Azdast T, Doniavi A (2013) An experimental study on mechanical properties of friction stir welded ABS sheets. Mater Des 43:402–409
15. Pirizadeh M et al (2014) Friction stir welding of thermoplastics using a newly designed tool. Mater Des 54:342–347
16. Mendes N et al (2014) Effect of friction stir welding parameters on morphology and strength of acrylonitrile butadiene styrene plate welds. Mater Des 58:457–464
17. Oliveira PHF et al (2010) Preliminary study on the feasibility of friction spot welding in PMMA. Mater Lett 64(19):2098–2101
18. Ahmadi H, Arab NBM, Ashenai Ghasemi F, Eslami Farsani R (2012) Influence of pin profile on quality of friction stir lap welds in carbon fiber reinforced polypropylene composite. Int J Mech Appl 2:24–28

19. Kiss Z, Czigány T (2007) Applicability of friction stir welding in polymeric materials. Polytech Mech Eng 51(1):15–18
20. Gao J et al (2014) Submerged friction stir weld of polyethylene sheets. J Appl Polym Sci 131(22)
21. Azarsa E, Mostafapour A (2014) Experimental investigation on flexural behavior of friction stir welded high density polyethylene sheets. J Manuf Process 16(1):149–155
22. Arici A, Sinmazçelýk T (2005) Effects of double passes of the tool on friction stir welding of polyethylene. J Mater Sci 40(12):3313–3316
23. Aydin M (2010) Effects of welding parameters and pre-heating on the friction stir welding of UHMW-polyethylene. Polym Plast Technol Eng 49(6):595–601
24. Liu FC, Liao J, Nakata K (2014) Joining of metal to plastic using friction lap welding. Mater Des (1980–2015). 54:236–244
25. Zafar A et al (2016) Investigating friction stir welding on thick nylon 6 plates. Weld J 95 (6):210S–218S
26. Husain IM et al (2015) Mechanical properties of friction-stir-welded polyamide sheets. Int J Mech Mater Eng 10(1):1–8
27. Paoletti A, Lambiase F, Di Ilio A (2015) Optimization of friction stir welding of thermoplastics. Procedia CIRP 33:563–568
28. DebRoy T, Bhadeshia H (2013) Friction stir welding of dissimilar alloys—a perspective. Sci Technol Weld Joining 15:266–270
29. Dashatan SH et al (2013) Friction stir spot welding of dissimilar polymethyl methacrylate and acrylonitrile butadiene styrene sheets. Mater Des 45:135–141
30. Eslami S et al (2015) Effect of friction stir welding parameters with newly developed tool for lap joint of dissimilar polymers. Procedia Eng 114:199–207
31. Bilici MK, Yukler AI (2012) Effects of welding parameters on friction stir spot welding of high density polyethylene sheets. Mater Des 33:545–550
32. Threadgill P et al (2009) Friction stir welding of aluminium alloys. Int Mater Rev 54(2):49–93
33. Bilici MK, Yükler Aİ (2012) Influence of tool geometry and process parameters on macrostructure and static strength in friction stir spot welded polyethylene sheets. Mater Des 33:145–152
34. Goushegir S, dos Santos J, Amancio-Filho S (2014) Friction spot joining of aluminum AA2024/carbon-fiber reinforced poly (phenylene sulfide) composite single lap joints: microstructure and mechanical performance. Mater Des 54:196–206
35. Mendes N et al (2014) A novel friction stir welding robotic platform: welding polymeric materials. Int J Adv Manuf Technol 1–10
36. Mostafapour A, Taghizad Asad F (2016) Investigations on joining of nylon 6 plates via novel method of heat assisted friction stir welding to find the optimum process parameters. Sci Technol Weld Joining 1–10
37. Vijendra B, Sharma A (2015) Induction heated tool assisted friction-stir welding (i-FSW): a novel hybrid process for joining of thermoplastics. J Manuf Process 20:234–244
38. Rui-Dong F et al (2011) Improvement of weld temperature distribution and mechanical properties of 7050 aluminum alloy butt joints by submerged friction stir welding. Mater Des 32(10):4825–4831
39. Kiss Z, Czigany T (2012) Microscopic analysis of the morphology of seams in friction stir welded polypropylene. Express Polym Lett 6(1):54–62
40. Colligan K (1999) Material flow behavior during friction welding of aluminum. Weld J 75 (7):229s–237s
41. Di S et al (2007) The influence of zigzag-curve defect on the fatigue properties of friction stir welds in 7075-T6 Al alloy. Mater Chem Phys 104(2):244–248
42. Rotman A (2009) Verification of the influence of surface energies on the effective mobility. Technion–Israel Institute of Technology
43. Song F, Wang Q, Wang T (2016) The effects of crystallinity on the mechanical properties and the limiting PV (pressure × velocity) value of PTFE. Tribol Int 93, Part A:1–10

44. Wang D et al (2014) Tribological evaluation of surface modified H13 tool steel in warm forming of Ti–6Al–4 V titanium alloy sheet. Chin J Aeronaut 27(4):1002–1009
45. Starkweather HW et al (1956) Effect of crystallinity on the properties of nylons. J Polym Sci 21(98):189–204
46. Sichina W (2000) DSC as problem solving tool: measurement of percent crystallinity of thermoplastics. Thermal Analysis Application Note
47. Ghasem N, Al-Marzouqi M, Duaidar A (2011) Effect of quenching temperature on the performance of poly(vinylidene fluoride) microporous hollow fiber membranes fabricated via thermally induced phase separation technique on the removal of CO_2 from CO_2-gas mixture. Int J Greenhouse Gas Control 5(6):1550–1558
48. Millot C et al. Assessment of polyamide-6 crystallinity by DSC. J Therm Anal Calorim 1–8
49. Leitão C et al (2008) Material flow in friction stir welding. Microsc Microanal 14(S3):87–90
50. Simões F, Rodrigues DM (2014) Material flow and thermo-mechanical conditions during friction stir welding of polymers: literature review, experimental results and empirical analysis. Mater Des 59:344–351
51. Arbegast WJ (2008) A flow-partitioned deformation zone model for defect formation during friction stir welding. Scripta Mater 58(5):372–376
52. Zafar A et al (2006) Visual analysis of material flow during friction stir welding of nylon-6. ARPN J Eng Appl Sci 11:13309–13312

Variation of Tool Offsets and Its Influence on Mechanical Properties of Dissimilar Friction Stir Welding of Aluminum Alloy 6061 and S235JR Mild Steel by Conventional Belting Milling Machine

W.M. Syafiq, M. Afendi, R. Daud, M.N. Mazlee and N.A. Jaafar

Abstract The tool offset parameter's influence on joint strength of friction stir welding (FSW) of aluminum alloy 6061 and S235JR steel performed on a conventional belting milling machine was evaluated. The type of welding defects that arose at different tool offset values and how they affect the ultimate tensile strength (UTS) of the joints produced are also studied. It was found that at +0.2 and 0 mm offset, welding was incomplete and tunnel defects were formed, with average tensile strengths of 22.91 and 41.62 MPa for joints welded at each tool offset values respectively. Tool offset value of −0.2 mm was found to produce the best joints in terms of tensile strength, at an average of 95.73 MPa. For all joints, in terms of defects it was found that incomplete welding severely reduced tensile strength of joints, while tunnel defects also deteriorated tensile strength of joints albeit to a lesser degree.

1 Introduction

Friction stir welding (FSW) is a welding technique that was first introduced by The Welding Institute (TWI), London in 1991 [1]. Its solid-state nature means it offers a few improvements over other conventional welding techniques, the most notable one being the reduction of heat input. Several solid-state bonding techniques have

W.M. Syafiq (✉) · M. Afendi · R. Daud
School of Mechatronic Engineering, Pauh Putra Campus, Universiti Malaysia Perlis (UniMAP), 02600 Arau, Perlis, Malaysia
e-mail: wmsyafiq10@gmail.com

M.N. Mazlee
School of Materials Engineering, Centre of Excellence for Frontier Materials Research, Universiti Malaysia Perlis, 01000 Kangar, Perlis, Malaysia

N.A. Jaafar
Advanced Technology Training Center (ADTEC), Lot 635, Jln Mahang, 09500 Karangan, Kulim, Kedah, Malaysia

M. Awang (ed.), *2nd International Conference on Mechanical, Manufacturing and Process Plant Engineering*, Lecture Notes in Mechanical Engineering,
DOI 10.1007/978-981-10-4232-4_3

gained interest because of this, such as ultrasonic joining, friction welding and diffusion bonding [2–4]. However, friction welding is limited because at least one-half of the joint must be circular. Diffusion welding is a long joining process, while ultrasonic welding is only applicable on thin plates. Compared to FSW, in addition to its solid-state nature it is also a relatively quick process and is able to weld different thicknesses and joint shapes such as lap joints and T-joints [5, 6]. Honda Motor Co. Ltd. incorporated FSW in their fabrication of the Accord 2013 and took advantage of the technique's various perks to weld the front sub-frame of the model, subsequently making Honda the first company in the automotive industry to apply the technique in mass production [7].

One important consideration when welding hard materials like steel is tool durability, as the tool gets severely damaged due to friction with steel and the resulting high temperature. Kimapong et al. found that for FSW of AA5083 and SS400 steel, a tool that was offset 0.2 mm into the steel produced joints with the best tensile strength [8]. Another important finding of the study was that sound joints were only produced when aluminum alloy was placed on the retreating side, due to the plasticized aluminum alloy being pushed to the activated faying surface of the steel. Chen and Kovacevic demonstrated that in FSW of AA6061 and AISI 1018 steel, when the tool was offset such that the pin only plunged into one side of the joint, an unsatisfactory joint was produced as the weld was generated only by the tool shoulder action [9]. Satisfactory joints were produced when the tool was placed with a 1 mm offset towards aluminum alloy, although the tool was found to be broken after travelling 100 mm from initial position. It is to note that in this study, steel was placed on the retreating side, instead of aluminum alloy. This is in contrast to the findings stated earlier by Kimapong et al., therefore it could be said that the placement of steel in the retreating side contributed to the weak joint produced by Chen and Kovacevic in their study.

More recently, Ramachandran et al. showed that in FSW of AA5052-H32 and HSLA steel, the tool offset that produced the strongest joint is 0.25 mm, i.e. the pin penetrated 0.25 mm into the steel side [10]. Joint tensile strength declined steadily as the tool pin was pushed further into the steel side, whereas the joint produced at tool offset of 0.1 mm was also weaker. A study by Karimi et al. on FSW of AA1100-H16 and AISI 1045 carbon steel also presented the importance of tool offset and tool material [11]. It was seen that tools made of high speed steel (HSS) got severely damaged i.e. tool pin eliminated after welding with tool offset of 3 mm towards steel side, whereas tools made of tungsten carbide (WC) experienced high tool wear by the end of weld. At tool offset of 1 mm into steel side, both tool materials experienced acceptable tool wear. It is clear that in performing FSW of aluminum alloy and steel, the placement of material in the retreating side and tool offset are two paramount points that need to be taken into consideration.

The aim of this paper is to investigate the effect of tool offset on the joint strength of FSW of AA6061 and S235JR steel performed on a conventional belting milling machine. The use of a conventional belting milling machine is chosen because it is easily available and a common piece of machinery in a workshop. Satisfactory FSW performance would mean that any workshop, industrial or academic, with a conventional belting milling machine will have the ability to weld aluminum alloy and steel. The effect of tool offset on joint's tensile strength was investigated by tensile testing. The formation of defects resulted from each tool offset was also studied.

2 Experimental Procedure

Aluminum alloy 6061 and S235JR mild steel plates of thickness 5 mm were welded using FSW technique on a conventional belting milling machine (MASTIKA B-3HS) [12]. The tool material was made of H13 tool steel with shoulder diameter of 20 mm and cylindrical pin diameter of 5 mm (refer Fig. 1). The faying surface of the plates were cleaned with acetone before welding to get rid of any grease. The parameters used in this study were tool rotational speed of 1100 rpm, 46 mm/min travel speed and 3° tool tilt.

In this study, tool offset was defined such that a negative value is assigned when the tool pin penetrates the steel side, i.e. at tool offset value of 0 mm, the pin only barely touches the steel faying surface (refer Fig. 2). Welded plates were cut to tensile testing specimens according to ASTM E8/E8 M-13a [13] and were tested on a universal testing machine (Shimadzu TCE-N300) [14] with a crosshead speed of 1 mm/min. Vickers microhardness test was done on a cross section perpendicular to the weld direction at three different depths i.e. 0.5 mm below the top surface, 0.5 mm above the bottom surface and at the middle. The indenter was a square-based diamond pyramid and was applied with a load of 200 N with dwell time of 15 s.

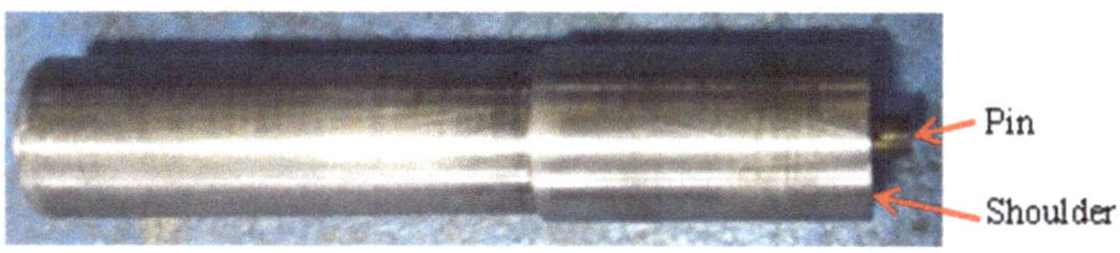

Fig. 1 FSW tool made of H13 tool steel

Fig. 2 Tool offset definition, with figure above depicting a tool offset value of 0 mm

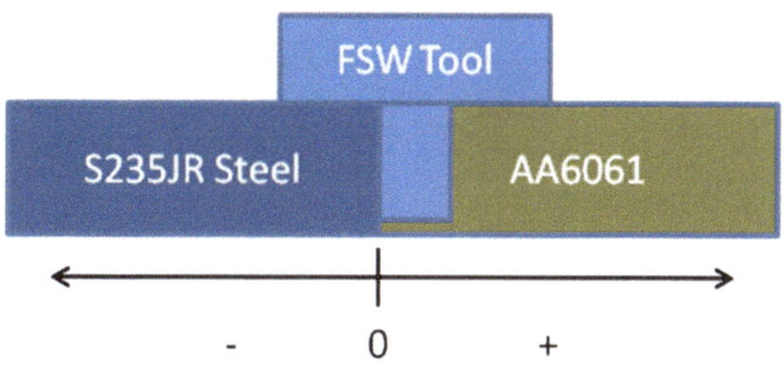

3 Results

3.1 *Effect of Tool Offset*

3.1.1 Tool Offset +0.2 mm

From Fig. 3, it is seen that tunnel defects were present in tensile specimens T1 and T2, with both tunnels about 1.2 mm in depth. It can be said that with a tool offset value of +0.2 mm, the weakest link of the FSW joint was located at the thin aluminum alloy layer at the faying surface. Both samples were very weak as they recorded low failure strength of 30.92 and 14.90 MPa respectively.

Figure 4 shows the failed surface of steel for T1 and T2 respectively. Aluminum alloy residue was seen only at the upper half of the surface, while the lower half

Fig. 3 Tunnel defects and thin aluminum layer at faying surface in **a** T1 and **b** T2

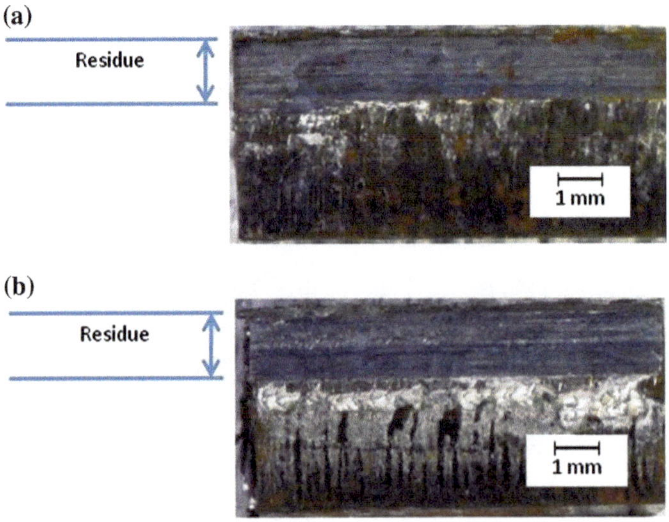

Fig. 4 Steel faying surface of **a** T1 and **b** T2 after failure

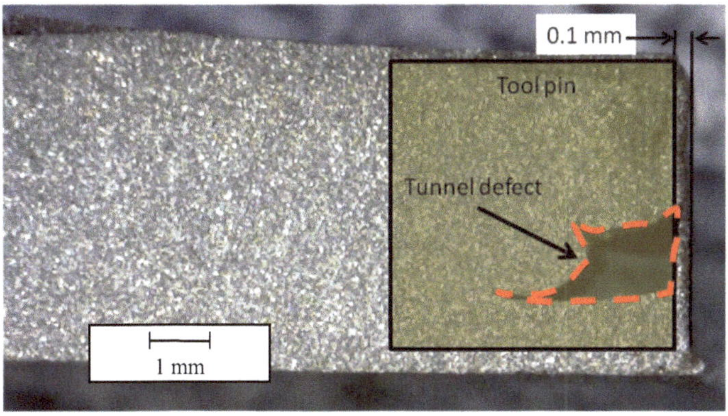

Fig. 5 Depiction of tool pin's passing area, tunnel defect position and thin aluminum layer on T1

remained relatively clean. It can be said that at tool offset of +0.2 mm, the plasticized aluminum alloy at the lower half of faying surface failed to sufficiently bond with steel. The extra thermomechanical action of the tool shoulder caused the upper half of aluminum alloy to sufficiently bond with steel.

Figure 5 shows the tool pin's passing area during FSW and tunnel defect. It is seen that the tunnel defect formed about 1 mm above tool pin's bottom surface. It is known that there is upwards material flow originating from under the tool pin during FSW [15]. Another important feature to be noted is the presence of a very thin layer (0.1 mm) of aluminum adjacent to the tunnel defect formation (in Fig. 5, this layer is located to the right of tunnel defect), and this was where the failure of the specimens happened. This thin aluminum layer can be attributed to the +0.2 mm tool offset. The plasticized aluminum alloy was merely pushed against the steel faying surface, without the steel faying surface actually interacting with the rotating tool pin. In this tool offset configuration, the steel faying surface was deemed 'inactive' as it did not come into direct contact with the tool pin that was plunged into the aluminum alloy [8].

3.1.2 Tool Offset 0 mm

Figure 6b, c show the failure mode of T3 and T4. It is seen that T3 failed at tunnel defect, while T4 failed at the faying layer, similar to the failure mode shown at tool offset +0.2 mm. Figure 6d shows the steel faying surface of T4. Similar to specimens T1 and T2, the welding here is also seen to be incomplete as only the upper half of the steel faying surface was welded to aluminum, although the failure strength of T4 was slightly higher than specimens welded with tool offset of +0.2 mm. The failure strength of T3 and T4 was recorded at 45.87 and 37.36 MPa respectively. Figure 6a shows the tunnel defect present in tensile specimen T3. The tunnel defect seen in T3 had a depth of 1 mm. From Fig. 6b, it is seen that the crack

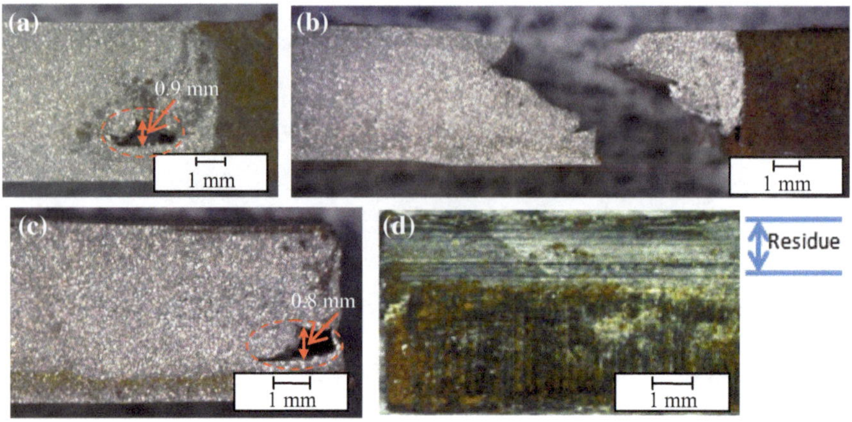

Fig. 6 Tensile specimen side views **a** T3 before failure, **b** T3 after failure, **c** T4 after failure and **d** steel fracture surface view of T4 after failure

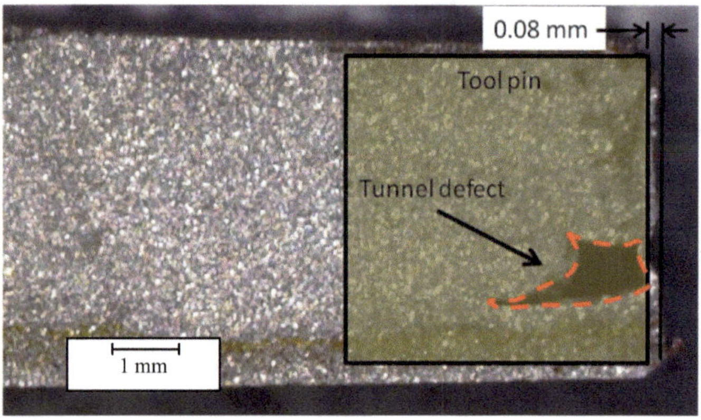

Fig. 7 Depiction of tool pin's passing area, tunnel defect and thin aluminum layer on T4

began at the root of the weld, and then propagated away towards the tunnel defect. It can be said that the crack originated at the root due to it being the weakest part of the joint, then travelled towards the tunnel defect which was the next weak link. Figure 7 shows the tool pin passing area of T4, tunnel defect with depth of 0.8 mm and thin aluminum layer of thickness 0.08 mm at the faying surface, which is thinner than the layer found on T1. In this specimen, the weakest link was found to be at the aluminum alloy layer.

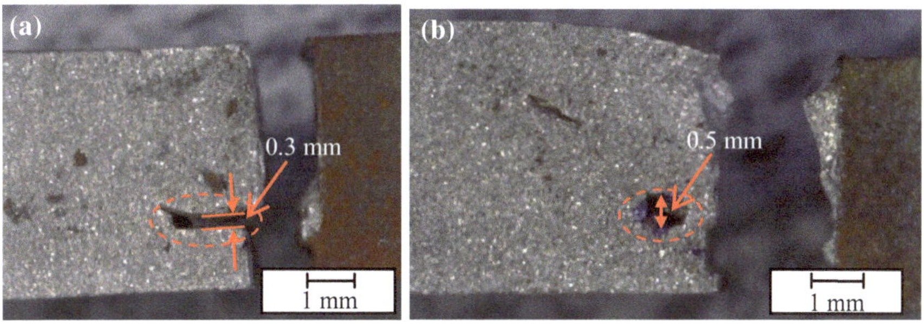

Fig. 8 Tensile specimens **a** T5 and **b** T7 after failure

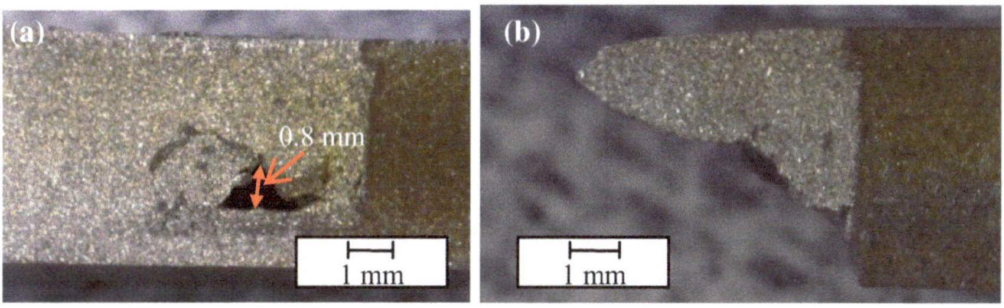

Fig. 9 Tensile specimen T6 **a** before and **b** after testing

3.1.3 Tool Offset −0.2 mm

Figure 8 shows tensile specimens T5 and T7 after recording joint strengths of 113.39 and 105.42 MPa respectively. Failure location in these two specimens was not influenced by the tunnel's presence, as it occurred at the interface. The tunnel depths of T5 and T7 were 0.3 and 0.5 mm respectively and were smaller than the 1 mm deep tunnel defect found in T3, which influenced the sample's fracture behavior. With these observations, it can be said that tunnel defects smaller than 0.6 mm no longer serve as the weakest link in the joint.

Figure 9 shows specimen T6 before and after tensile testing. Figure 9a displays a tunnel defect of 0.8 mm depth. As can be seen in the Fig. 9b, the presence of tunnel defect affected the fracture path of the specimen. The tunnel defect seen in T6 was also the largest out of all the tunnel defects seen in specimens in this category, which seems to relate to its low joint strength of 68.37 MPa. This finding seems to be in agreement with the previous hypothesis that with this experimental setup, tunnel defects with depths under 0.6 mm do not influence the failure location and joint strength. Results of the effect of tool offset on joint are summarized in Table 1.

Table 1 Summary of results

Tool offset (mm)	Specimen	Ultimate tensile strength (MPa)	Tunnel depth (mm)	Failure location
+0.2	T1	29.82	1.25	Aluminum alloy layer
+0.2	T2	14.90	1.20	Aluminum alloy layer
0.0	T3	45.87	0.90	Root to Tunnel
0.0	T4	37.36	0.80	Layer
−0.2	T5	113.39	0.30	Interface
−0.2	T6	68.37	0.80	Root to Tunnel
−0.2	T7	105.42	0.50	Interface

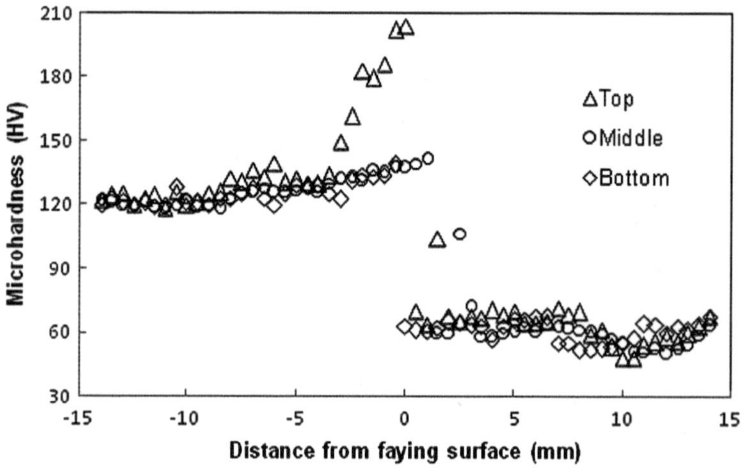

Fig. 10 Vickers Microhardness profiles taken perpendicular to welding direction at three different depths

3.2 *Microhardness*

Figure 10 shows the Vickers microhardness (HV) profile taken close to the top surface, at mid thickness and close to the bottom surface. The pattern seen in this figure was observed in all samples. The results shown here was taken from sample T5. It is seen that the microhardness value fluctuates in the nugget zone (NZ). This was caused by the scattering of steel particles separated by the interaction of rotating pin and steel faying surface during welding, as well as the presence of intermetallics in the nugget zone [16]. Hardness value in the steel side began to rise as it got closer to the nugget. In the base metal region, a hardness value of around 124 HV was obtained for all three profiles. As it got closer to the area worked by the FSW tool shoulder, the hardness value increased for the three profiles.

This observation is indicative of work hardening of the mild steel surface by the FSW tool shoulder [17]. It is seen that the hardness of the top surface profile is higher than the other two profiles, reaching a maximum of 204.1 HV. This phenomenon is similar to previous studies on hardness of FSW of aluminum and steel [18]. The mid thickness and bottom profiles both had maximum hardness values of around 140 HV. This can be attributed to less degree of work hardening experienced by areas away from the tool shoulder.

The hardness value was seen to be at the minimum at the aluminum alloy's thermomechanically affected zone (TMAZ) and heat affected zone (HAZ) border. During FSW, the thermomechanical conditions of the process is known to dissolve the hardening second phase particles in precipitation hardened aluminum alloys, which causes softening of the welded joint [18]. The grain size of the NZ is well known to be finer than the grains in TMAZ and HAZ due to dynamic recrystallization [19], and this could explain the observed phenomena by looking at the Hall-Petch hardness relation [20]:

$$Hv = H_0 + k_H d^{-1/2} \tag{1}$$

where Hv is hardness, d is grain size and H_0 and k_H are appropriate constants associated to measurements of hardness. The Hall-Petch relation states that the hardness of a material increases as its grain size decreases, which explains why the minimum hardness value was located in TMAZ/HAZ border. This pattern of hardness value recorded is similar to most FSW joints that involve precipitation hardened aluminum alloys like AA6XXX [21].

The minimum hardness for top, mid thickness and bottom profiles were recorded at 48, 50.7 and 51.5 HV at distances 10.5, 10.5 and 8 mm from the faying surface. The bottom profile was seen to record a minimum hardness value closer to the faying surface than the other profiles, which indicates the TMAZ/HAZ border is closer. This is due to the narrowing of TMAZ away from the top surface due to less thermomechanical action by the tool shoulder reaching the lower surface [18, 22].

4 Conclusions

This paper investigates the performance of conventional belting milling machine to perform FSW and the influence of tool offset on the joint performance of friction stir welded aluminum alloy AA6061 to S235JR mild steel. The tool offset values used in experiment were +0.2, 0 and −0.2 mm. The conclusions obtained from the investigations are presented below.

- Out of all the tool offset values used, the tool offset −0.2 mm (tool pin slightly touches steel) performed the best at an average tensile strength of 95.73 MPa, compared to tool offset values 0 and +0.2 mm with average tensile strengths of 41.62 and 22.91 MPa respectively.

- At tool offset of +0.2 mm, welding was found to be incomplete based on observations on failure surfaces, as only upper halves of the faying surfaces were successfully welded. The incomplete welding resulted in weak joining at the faying surface, and it served as the weakest link in the joint even with the presence of tunnel defects of depth 1.2 mm as failure did not occur at the tunnel.
- At tool offset of 0 mm, the failure mode depended on the size (depth) of tunnels and the height of successfully welded faying surfaces. At tunnel depth of 1 mm and successfully welded surface height of 3.9 mm, the crack propagated through the tunnel. On the other hand, failure occurred at the faying surface at tunnel depth of 0.8 mm and successfully welded surface height of 2.3 mm.
- The hardness value of the FSW joints was found to be at the minimum at the aluminum alloy TMAZ/HAZ border due to the softening effect caused by dissolution of hardening precipitates of AA6061 as well as larger grain size compared to the NZ. The hardness profile obtained was similar to other FSW joints involving precipitation hardened aluminum alloys.
- Creating a dissimilar joint of aluminum alloy and steel is possible by a commonly available conventional belting milling machine by means of FSW. However, further study must be done on the defect formations in order to improve joint performance.

References

1. Thomas W (1991) Friction stir butt welding. International Patent Application No. PCT/G892/0220
2. Tsujino J, Hidai K, Hasegawa A et al (2002) Ultrasonic butt welding of aluminum, aluminum alloy and stainless steel plate specimens. Ultrasonics 40:371–374
3. Yamamoto N, Takahashi M, Aritoshi M, Ikeuchi K (2005) Effect of interfacial layer on bond strength of friction-welded interface of Al-Mg5083 alloy to mild steel. Q J Jpn Weld Soc 23:496–503. doi:10.2207/qjjws.23.496
4. Kuroda S, Saida K, Nishimoto K (1999) Microstructures and properties of directly joint of A6061 aluminum alloy and SUS316 stainless steel. Q J Jpn Weld Soc 17–3:484–489
5. Kimapong K, Watanabe T (2005) Lap joint of A5083 aluminum alloy and SS400 steel by friction stir welding. Mater Trans 46:835–841
6. Fratini L, Buffa G, Filice L, Gagliardi F (2006) Friction stir welding of AA6082-T6 T-joints: process engineering and performance measurement. Proc Inst Mech Eng Part B J Eng Manuf 220:669–676. doi:10.1243/09544054JEM327
7. Kusuda Y (2013) Feature article Honda develops robotized FSW technology to weld steel and aluminum and applied it to a mass-production vehicle. Ind Robot Int J 40:208–212. doi:10.1108/01439911311309889
8. Kimapong K, Watanabe T (2004) Friction stir welding of aluminum alloy to steel. Weld J 83:277–282
9. Chen CM, Kovacevic R (2004) Joining of Al 6061 alloy to AISI 1018 steel by combined effects of fusion and solid state welding. Int J Mach Tools Manuf 44:1205–1214. doi:10.1016/j.ijmachtools.2004.03.011

10. Ramachandran KK, Murugan N, Kumar SS (2015) Effect of tool axis offset and geometry of tool pin profile on the characteristics of friction stir welded dissimilar joints of aluminum alloy AA5052 and HSLA steel. Mater Sci Eng A 639:219–233. doi:10.1016/j.msea.2015.04.089
11. Karimi N, Nourouzi S, Shakeri M et al (2012) Effect of tool material and offset on friction stir welding of Al alloy to carbon steel. Adv Mater Res 445:747–752. doi:10.4028/www.scientific.net/AMR.445.747
12. Hasil Karya (2014) MASTIKA B-3HS milling machine catalogue http://www.hasilkarya.com.my/ecatalog/MILLING-MACHINE/3HP-Milling/B3.pdf
13. ASTM International (2009) Standard test methods for tension testing of metallic materials 1. Astm 1–27. doi:10.1520/E0008
14. SHIMADZU (Shimadzu Corporation) (2016) Universal testing/tensile testing machine: SHIMADZU (Shimadzu Corporation). http://shimadzu.com/an/test/universal/index.html
15. Asadi P, Akbari M, Karimi-Nemch H (2014) 12—Simulation of friction stir welding and processing. In: Givi MKB, Asadi P (eds) Advances in friction stir welding processing. Woodhead Publishing, p 511
16. Kundu S, Roy D, Bhola R et al (2013) Microstructure and tensile strength of friction stir welded joints between interstitial free steel and commercially pure aluminium. Mater Des 50:370–375. doi:10.1016/j.matdes.2013.02.017
17. Chen TP, Lin W-B (2010) Optimal FSW process parameters for interface and welded zone toughness of dissimilar aluminium–steel joint. Sci Technol Weld Join 15:279–285. doi:10.1179/136217109X12518083193711
18. Uzun H, Dalle Donne C, Argagnotto A et al (2005) Friction stir welding of dissimilar Al 6013-T4 to X5CrNi18-10 stainless steel. Mater Des 26:41–46. doi:10.1016/j.matdes.2004.04.002
19. Krasnowski K, Hamilton C, Dymek S (2015) Influence of the tool shape and weld configuration on microstructure and mechanical properties of the Al 6082 alloy FSW joints. Arch Civ Mech Eng 15:133–141. doi:10.1016/j.acme.2014.02.001
20. Sato YS, Urata M, Kokawa H, Ikeda K (2003) Hall-Petch relationship in friction stir welds of equal channel angular-pressed aluminium alloys. Mater Sci Eng A 354:298–305. doi:10.1016/S0921-5093(03)00008-X
21. Threadgill PL, Leonard AJ, Shercliff HR, Withers PJ (2009) Friction stir welding of aluminium alloys. Int Mater Rev 54:49–93. doi:10.1179/174328009X411136
22. Mishra RS, Ma ZY (2005) Friction stir welding and processing. Mater Sci Eng R Rep 50:1–78. doi:10.1016/j.mser.2005.07.001

Fracture Behavior of Intermetallic Compound (IMC) of Solder Joints Based on Finite Elements' Simulation Result

E.P. Ooi, R. Daud, N.A.M. Amin, F. Mat, M.H. Sulaiman, M.S. Abdul Majid, M. Afendi and A.K. Ariffin

Abstract The development of microelectronic industry has made solder joints failure a major reliability issue. From literature, many researchers have identified that intermetallic compounds (IMC) layer contribute greatly to the fracture of solder joint. This paper presents a finite element modeling of solder butt joints IMC layer failure based on displacement extrapolation method (DEM). Conceptual study on single edge crack of IMC solder joints is presented. A FE analytical model is proposed to be used in difference range of crack length to understand the fracture behavior of solder joint of IMC layer. The simulation results show that soldering material become less tough if greater crack length is present in the joint. It also seen that the thicker IMC has slightly reduced the stress intensity factor on the crack tip but the change from solder to IMC layer decrease the solder joint fracture toughness.

1 Introduction

One of the major factor cause microelectronic products failure is solder joint fracture due to thermo-mechanical stress. Recently, many of the intermetallic compounds (IMCs) were found to grow on the interface between the pad and the solder material [1]. The brittle property of IMCs caused the strong stress concentration effect during the thermal cycles [2, 3] and mechanical impacts [1, 4–10]. The cracks were found to initiate and propagate near the IMC layers. Thus, the presence of IMC has greatly influence the solder joints reliability and many studies [1–6, 8, 9] were based on it.

E.P. Ooi (✉) · R. Daud · N.A.M. Amin · F. Mat · M.H. Sulaiman · M.S. Abdul Majid · M. Afendi
Fracture and Damage Mechanic Research Group, School of Mechatronic Engineering, University Malaysia Perlis, 02600 Arau, Perlis, Malaysia
e-mail: eangpang@yahoo.com.my

A.K. Ariffin
Faculty of Engineering and Built Environment, Universiti Kebangsaan Malaysia, 43600 UKM Bangi, Selangor, Malaysia

M. Awang (ed.), *2nd International Conference on Mechanical, Manufacturing and Process Plant Engineering*, Lecture Notes in Mechanical Engineering,
DOI 10.1007/978-981-10-4232-4_4

There are few factors that influence fracture toughness of solder joint, such as solder joint thickness and crack location related to interface of solder joint. Alam et al. [4] have showed that the stress intensity factors (SIF) (e.g., K_I and K_{II}) increase with the thickness of solder alloys at the loading rate of 0.8 MPa/s for the crack located at the middle of IMC layer. Alam [4] also highlighted that the nearer the crack location to the interface, the higher the SIF values (both K_I and K_{II}). The K_{II} increases to a very high value that could lead to a very unstable crack due to its higher creep resistance. The simulation result proves that the crack nearer the interface is always having more tendencies to propagate.

Yoa et al. [6] has developed a 3D finite element solder interconnect model using the unified creep plasticity (UCP) theory and cohesive zone model (CZM). The result revealed that the von Mises stress needed to initiate a crack become smaller for the thicker intermetallic layer. The solder joint lifetime decreases with IMC layer thickness. However, according to Alam et al. [4], the thinner IMC layers have strong effect on the SIF which leading to unstable crack.

In relation to this, Liu [8] and Shen & Alulu [9, 10] had analysed and determined the fracture mechanic parameters (SIF and G value) for the elastic IMC layer on the effect of mechanical impact (shear rate on mode I and II). Due to the two difference results obtained by Siva and Jen [2], Alam [4] and Yao [6] on the IMC thickness effect, the mechanical loading on numerical modeling need to be reinvestigated on IMC thickness effect on stress failure.

Therefore, this paper aim to investigate the error correction of FE model and analytical model. The effect of crack length, solder thickness, IMC layer thickness and Young's Modulus on the fracture behaviour of solder joint are carried out by using linear elastic fracture mechanics (LEFM) approach to measure the stress intensity factor.

2 Methods

A validation study is conducted to find an optimum geometry for a numerical model based on Brown & Srawley analytical equation. According to Brown & Srawley analytical equation, C, the SIF value is calculated as:

$$K_{BS} = C\sigma\sqrt{\pi a} \tag{1}$$

$$C = 1.12 - 0.23\left(\frac{a}{w}\right) + 10.6\left(\frac{a}{w}\right)^2 - 21.7\left(\frac{a}{w}\right)^3 + 30.4\left(\frac{a}{w}\right)^4 \tag{2}$$

where K_{BS} = SIF for Brown & Srawley model, C = single Edge Correction Factor (constant), σ = loading applied, a = crack length and w = width of solder.

A finite element (FE) test model with geometry height 3.0 mm and width 2.0 mm and IMC thickness of 0.010 mm is setup as shown in Fig. 1. A crack with

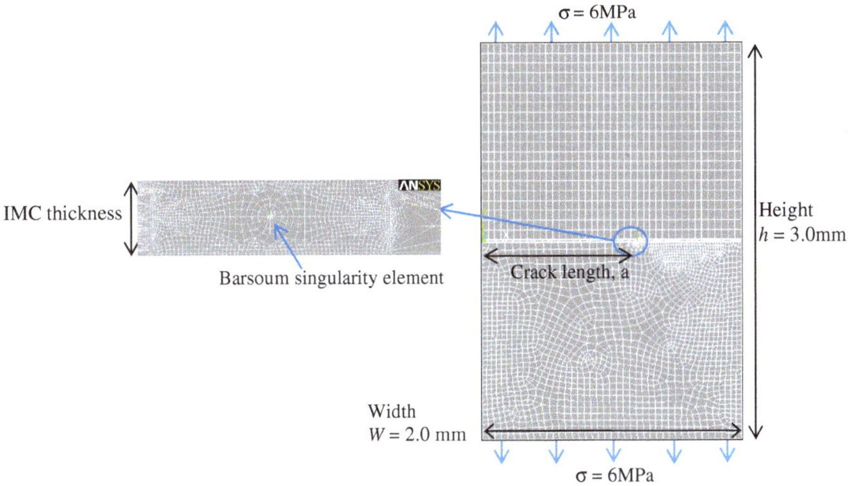

Fig. 1 FE Model on Brown & Srawley Analytical equation

Table 1 Material properties for the solder joint modeling

Materials	Elastic modulus, E (GPa)	Poisson ratio, v
Intermetallic compound(IMC)	110	0.30
Lead-free soldeer SnAgCu (SAC)	43	0.30
Leaded solder Sn37Pb Eutectic (SnPb)	35	0.36
Copper (Cu)	129	0.34

length of 0.2 mm–1.8 mm is introduced to study the stress intensity factor at the crack tip where a tensile stress σ of 6 MPa is applied on both upper and bottom surfaces.

2.1 FE Solder Joint Modeling

In order to simulate and characterize the typical solder joint, one solder joint FE model which validated with Brown & Srawley analytical equation have been developed. The FE analysis has been conducted using ANSYS Release13. FE model involved three types of materials—copper, lead-free tin-silver-copper SnAgCu and intermetallic compound IMC. The meshing scheme was set to isoperimetric quadrilateral elements (PLANE 183) with Barsoum singularity element around the crack tip. Convergence study was conducted to confirm the accuracy of meshing. Table 1 shows the material properties of solder joints. Copper and IMC layer are isotropic linear elastic materials while solder matter is a visco-plastic material. This analysis is based on static loading without applying loading rate and thermal effect. A crack is introduced at the left middle edge of IMC layer in order to measure the SIF at its crack tip.

A finite element (FE) test model with geometry height 3.0 mm and width 2.0 mm and IMC thickness of 0.010 mm is setup as shown in Fig. 1. A crack with length of 0.2 mm–1.8 mm is introduced to study the stress intensity factor at the crack tip where a tensile stress σ of 6 MPa is applied on both upper and bottom surfaces.

First FE model of solder joint is setup as depicted in Fig. 2. Difference shear stress is applied from 4 MPa to 40 MPa to study the effect of different crack length on fracture parameter (SIF). The model height is 3.0 mm while its copper thickness and solder thickness is set at 1.0 mm. Its width is 2.0 mm and its IMC thickness is 0.006 mm while the crack length is varies from 0.05 mm to 1.0 mm.

Second FE model manipulates IMC thickness at 0.006, 0.012, 0.020 and 0.100 mm while the model height and width remains the same. The copper thickness is set at 1.35 mm, solder thickness 0.3 mm and the crack length is fixed at 0.05 mm. Difference shear stresses are applied on the model to study the effect of IMC thickness on fracture parameter (SIF).

For third FE model, difference shear stresses are applied on the model to study the effect of different Young's Modulus on fracture parameter (SIF). Model height and width remain as same geometry and Young's Modulus of IMC layer is set at 43, 110 and 129 GPa the rest of geometry model thickness, model width, copper thickness and solder thickness are same as second FE model except crack length is set at 0.012 mm and its IMC thickness is 0.006 mm.

According to Alam [4], the critical value for fracture toughness. (Kic) of Cu_6Sn_5 was in the range of 1.4 and 3 MPa$\sqrt{m}$. For this study, Kic = 1.4 MPa$\sqrt{m}$ is selected which crack would propagate when the K_I is more than Kic.

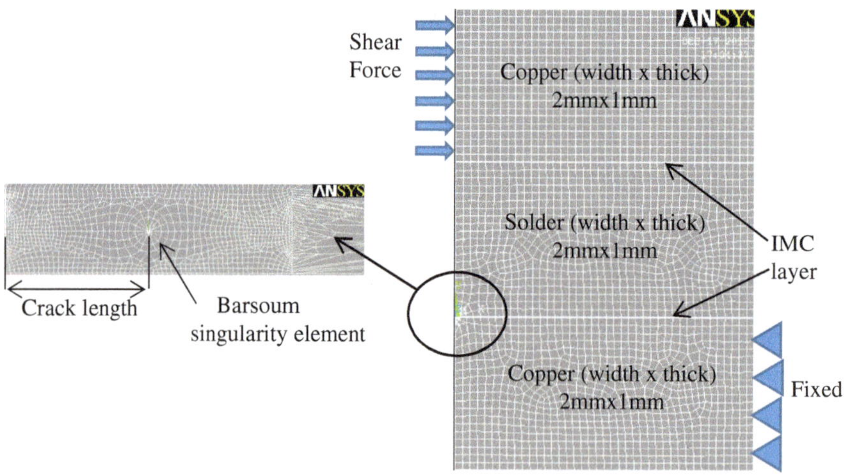

Fig. 2 The FE Model validated with Brown & Srawley Analytical equation for small crack length studies

3 Results and Discussion

3.1 Validation of FE Model with Brown & Srawley Equation

Figure 3 graphs shows the C factor of Brown & Srawley analytical equation and a FE model with only IMC material. The % C error is around 0.7% for *a/w* range from 0.1 to 0.6.

From the above validation study studies, the geometry of this FE model can be used to study the behaviour of solder joint material. In the present studies, length of crack introduced are start from edge toward the middle of IMC layer of the solder joint.

3.2 The Effect on Different Crack Length on Fracture Parameter

In the first FE model, a shear stress of 4–40 MPa were applied to the FE full model to study the crack length effect on solder joint fracture toughness, the SIF results of K_I and K_{II} are shown in Fig. 4a, b.

From the Fig. 4a, b, longer crack length will cause greater SIF for the same load applied. According to Fracture equation, SIF is directly proportional to square root of crack length ($K_{BS} = C\sigma\sqrt{\pi a}$). The strength of IMC layer of solder joint against load reduce when the size of microcrack (delamination or void) or crack length increase. Smaller load can overcome the Kc of the material and cause it to fracture. This result has shown agreement with Xu Chen [7] data for strain energy release rate which increase with normalised crack length of solder joint. Strain energy release rate, $G = K^2/E$ where K is SIF and E is Young's modulus. The graph also show that the fracture parameter (K_I and K_{II}) increase with the shear stress.

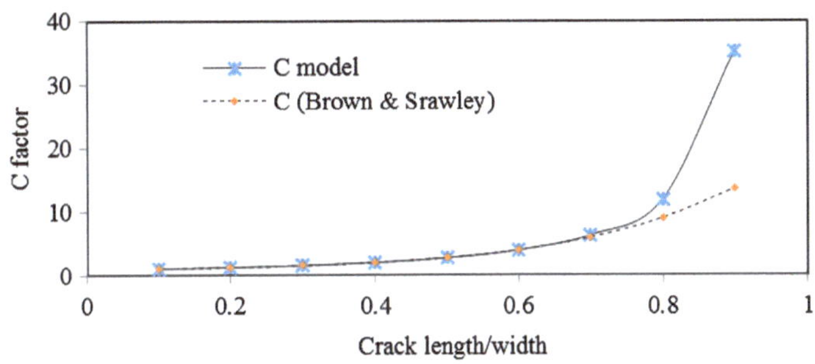

Fig. 3 Comparison between SIF computed from FE model and Brown & Srawley equation

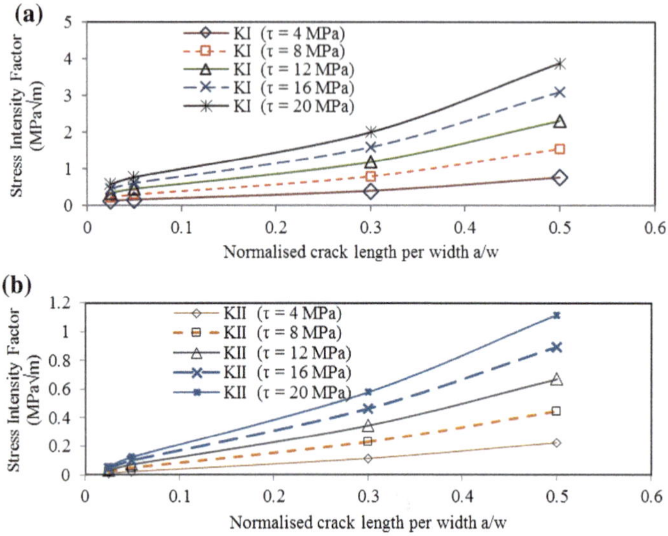

Fig. 4 **a** Stress intensity factor K_I obtained when difference shear stress applied on varying crack length. **b** Stress intensity factor K_{II} obtained when difference shear stress applied on varying crack length

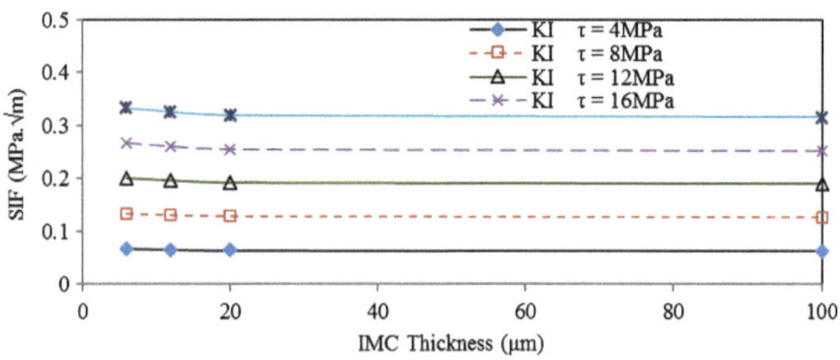

Fig. 5 The effect of difference IMC thickness on stress intensity factor K_I for varying shear stress

3.3 *The Effect on Different IMC Thickness on Fracture Parameter*

From Fig. 5, the increase of IMC thickness (from 6 μm to 100 μm) will cause a slight decrease in SIF value from 6 μm to 20 μm. No change in SIF for IMC thickness more 20 μm shows that IMC thickness doesn't affect the SIF if it is more than 20 μm. When the IMC thickness increase, the solder joint become slightly stronger and has resistance toward the mechanical stress. M.O. Alam [4] reported that a thicker IMC layer slightly reduces crack propagation propensity and Jen [2] simulation shown same result which the SIF value almost not affected by the thickness of IMC layer. (for thickness more than 20 μm) However, in practical the IMC thickness found to grow during reflow, thermal cycling and aging, and many researchers [2, 4, 6] have summarized that the thicker the IMC layer will make the solder joint fracture toughness to become lower. This opinion is contradicted to this simulation result but it can be explained when the influence of the material's Young's modulus on fracture toughness is discussed in following simulation model.

3.4 *The Effect on Different Young's Modulus on Fracture Parameter*

During soldering, the interaction and inter-diffusion reaction between the solder pad and copper metal forms Cu_6Sn_5 IMC which changes the properties of the solder joint. The Young's modulus of IMC is around 110 GPa as compared to SnAgCu solder 43 GPa. Thus, the grow of IMC layer has increased the modulus of the copper/solder interface.

From FE simulation on the Young modulus factor, the result in Fig. 6a shows that the Young's modulus of material has greatly influence the fracture parameter,

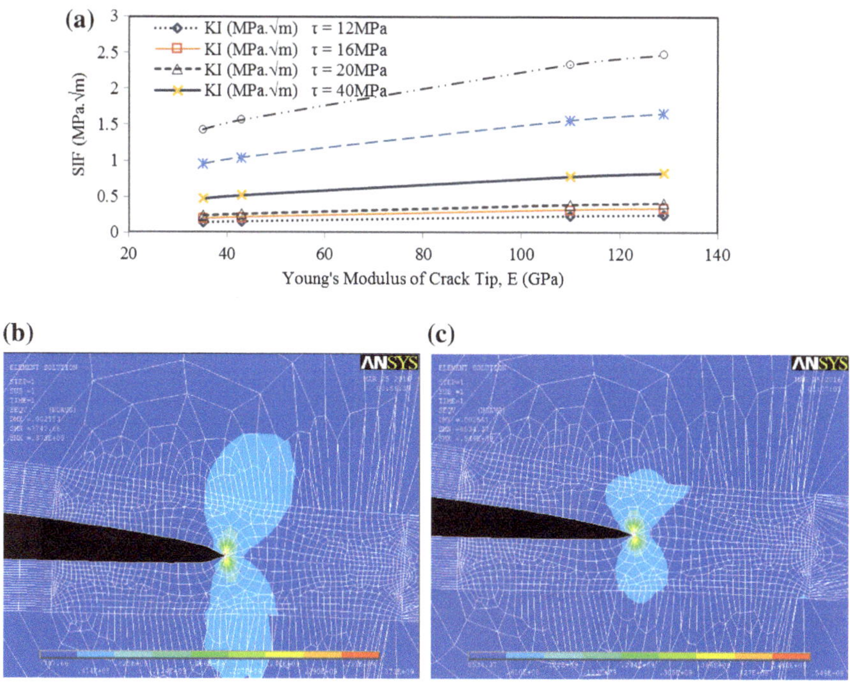

Fig. 6 **a** The effect of difference Young's Modulus on stress intensity factor K_I for varying shear stress. **b** and **c** von Misses stress distribution for crack tip Young's modulus 43 and 110 GPa respectively under shear stress of 12 MPa and crack length of 0.05 mm

SIF. The change of solder matter to IMC material has made the overall effect of IMC growth become significantly increase the SIF upon static stress and weaken the fracture toughness of the solder joint. The solder material becomes more brittle as the modulus increase and lead to crack nucleation and propagation. The stress intensity factor produced in solder material under 80 MPa shear stress is 1.04 $\text{MPa}\sqrt{\text{m}}$ and has increased to 1.56 $\text{MPa}\sqrt{\text{m}}$ when it turn into IMC. This change has exceeded its fracture toughness of 1.4 $\text{MPa}\sqrt{\text{m}}$. Thus, the effect of the IMC thickness become negligible when included the factor of Young's modulus as the overall simulation result.

Figure 6b and c reveal the von Misses stress distribution at the crack tip for Young's modulus 43 and 110 GPa respectively under the same shear stress of 12 MPa and crack length of 0.05 mm. This shows that the stress at the crack tip increase when the value of Young's modulus become greater. The numerical finding clearly suggests that the fracture toughness of the joint decrease with the IMC layer grows. Therefore, in daily life, material become fatigue and easily break due to the repeated thermal stress and aging process which turn most of the jointed material (like welding & soldering) become something brittle with higher Young modulus properties like IMC.

4 Conclusions

This simulation study has identified FE model geometry which validated with suitable analytical model use for difference range of crack length to understand the fracture behaviour of solder joint's IMC layer. The result shows that the material toughness against stress reduces when crack size or internal flaw increase. Finally, although the IMC thickness has slight impact to the stress intensity factor but the growth of IMC layer obviously increase its Young's modulus of the joint and cause the material to become brittle and crack more easily.

Acknowledgements The authors would like to be obliged to Fundamental Research Grant Scheme FRGS 9003-00491 and My Ph.D. from Ministry of High Education (MOHE) of Malaysia for providing Research Fund of Doctoral Program.

References

1. Tian Yanhong, Hang Chunjin, Wang Chunqing et al (2011) Effects of bump size on deformation and fracture behavior of Sn3. 0Ag0. 5Cu/Cu solder joints during shear testing. Mater Sci Eng A 529:468–478
2. Jen Yi-Ming, Chiou Yung-Chuan, Chen-Li Yu (2011) Fracture mechanics study on the intermetallic compound cracks for the solder joints of electronic packages. Eng Fail Anal 18:797–810

3. Amalu Emeka H, Ekere Ndy N (2012) High temperature reliability of lead-free solder joints in a flip chip assembly. J Mater Process Technol 212:471–483
4. Alam MO, Lu H, Bailey C, Chan YC (2009) Fracture mechanics analysis of solder joint intermetallic compounds in shear test. Comput Mater Sci 45:576–583
5. Nadimpalli Siva PV, Spelt Jan K (2011) Effect of geometry on the fracture behavior of lead-free solder joints. Eng Fract Mech 78:1169–1181
6. Yao Yao, Keer Leon M, Fine Morris E (2010) Modeling the failure of intermetallic/solder interfaces. Intermetallics 18:1603–1611
7. Chen X, Lin YC, Liu X et al (2005) Fracture mechanics analysis of the effect of substrate flexibility on solder joint reliability. Eng Fract Mech 72:2628–2646
8. Liu De-Shin, Kuo Chia-Yuan, Hsu Chang-Lin et al (2008) Failure mode analysis of lead-free solder joints under high speed impact testing. Mater Sci Eng A 494:196–202
9. Shen Y-L, Aluru K (2010) Numerical study of ductile failure morphology in solder joints under fast loading conditions. Microelectron Reliab 50:2059–2070
10. Aluru K, Wen F-L, Shen Y-L (2011) Direct simulation of fatigue failure in solder joints during cyclic shear. Mater Des 32:1940–1947

Effect of Isothermal Aging on Mechanical Properties of Sn-3.0Ag-0.5Cu Solder Alloy with Porous Cu Interlayer Addition

N.H. Jamadon, N.D. Ahmad, F. Yusof, T. Ariga, Y. Miyashita and M.H.A. Shukor

Abstract The performance of Pb-free SAC305 solder joint added with porous Cu interlayer was investigated. The porous Cu interlayer was placed in a sandwich-like layer between SAC305 solder alloy and rod Cu as the substrate metal. Two different pores, 15 ppi (pore per inch, P15) and 25 ppi (P25) with the pore size approximately ϕ1.7 and ϕ1.0 mm, respectively were used. The soldering process of solder joint was carried out in three different soldering temperatures of 267, 287 and 307 °C while the holding time was set at 300 s. Tensile test was performed to evaluate the joining strength of the solder alloy with loading rate of 0.5 mm/min. The highest joint strength was recorded at 53 MPa when soldered with P25 porous Cu interlayer at 307 °C. High thermal aging test was performed in an oven, heated to temperature of 150 °C for aging time of 100, 200 and 500 h to investigate the heat and time effect on the joint strength. The results indicate the reduction of strength at the aged sample with the increased aging time. The crack initiates dominantly at the interfaces between the porous Cu/SAC305. In general, solder joint soldered with P25 of porous Cu joints have greater tensile strength (23 MPa) than P15 of porous Cu interlayer after aging test.

1 Introduction

Power electronics has been used in aerospace, automotive and energy production industries. Most of these power electronics packaging requires the use of solders that can withstand elevated temperature and extreme environment conditions [1, 2]. The typical solder alloys used under such environments are usually Pb-based.

N.H. Jamadon (✉) · N.D. Ahmad · F. Yusof · T. Ariga · M.H.A. Shukor
Center of Advanced Manufacturing and Material Processing (AMMP Centre),
Department of Mechanical Engineering, Faculty of Engineering, University of Malaya,
50603 Kuala Lumpur, Malaysia
e-mail: nashrahhani@gmail.com

Y. Miyashita
Department of Mechanical Engineering, Nagaoka University of Technology, Nagaoka, Japan

M. Awang (ed.), *2nd International Conference on Mechanical, Manufacturing and Process Plant Engineering*, Lecture Notes in Mechanical Engineering,
DOI 10.1007/978-981-10-4232-4_5

However, in 2006, the restriction of Hazardous Substances (RoHS) issued a directive that limit the usage of Pb-contained solders and proposed a shift to Pb-free solders for application in the electronic industry [3]. The tin-based (Sn-based) Pb-free solders with high Ag content such as Sn-3.0Ag-0.5 wt% Cu (SAC305) is particularly promising candidate for replacement of toxic Pb-contained solders because of its superior mechanical and thermal properties [4, 5]. The liquidus point of SAC305 is 219 °C so that it requires a higher reflow temperature of at least 240 ° C. This makes it suitable for use in middle-temperature-ranged applications [6]. Its application in elevated temperature conditions ranging from 270 °C to 350 °C as required for some industrial purposes is not suitable since its melting point of 219 ° C falls far below the defined range [7, 8]. Several attempts have therefore been made to adapt the SAC305 alloy for use in high temperature regime.

Studies from Noh et al. showed that the rapid growth of intermetallic compound (IMC) was observed in SAC305 with addition of rare-earth (RE) element of cerium (Ce) [9]. By adding 0.2 wt% zinc (Zn) into SAC305 + Ce solder the IMC growth was inhibited and its joint strength improved. However, information on its reliability at thermal cycling is not available for Pb-free solders when added with RE element [10]. Adding a fourth element into the SAC305 composition has also been found to increase the joint reliability. For example, the addition of cobalt (Co) or cobalt-nickel (CoNi) into the SAC305 solder alloy have both been shown to improve solder quality by inhibiting the excessive formation of intermetallic compound (IMC) layer [11]. However, the authors did not specifically discuss the joint performance of these Sn-3.0Ag-0.5Cu-Co and Sn-3.0Ag-0.5Cu-CoNi solder alloys.

Another approach to improve the performance of the SAC305 solder joint is by modifying its joining process. In fact there are already studies which involve an interlayer structure to be added in joining technique for dissimilar bonding. For instance, Liu et al. reported the use of composite preform in high temperature transient liquid phase (TLP) bonding process for die attach bonding application has increased the mechanical properties and reduced the IMC growth [12]. T. Zaharinie et al. has also evaluated the addition of a porous metal interlayer sandwiched in between the brazing fillers to improve the direct joining of sapphire to inconel [13]. This promising technique has been successfully demonstrated in brazing process but has yet to be attempted in solder joining.

The objective of the present study is to observe the high temperature performance characteristics of Pb-free SAC305 solder joint by modifying the physical soldering process. This involves the incorporation of a porous Cu interlayer in the soldering configuration. Evaluation on the tensile strength and observation on the fracture surface of the solder joint are carried out. In addition, it is known that the microstructure of the soldered joints changed significantly when exposed to high temperature thereby influencing their joint strength. For this reason, the effect of thermal aging on the microstructure and mechanical properties of soldered joints will be reported.

2 Experimental Procedures

2.1 Material Selection and Sample Preparation

Sn-3.0Ag-0.5Cu (wt%, SAC305) of lead free solder paste (Nihon Handa Co. Ltd.) were used as the solder material in this experiment to join 2 cylindrical 99.9% pure Cu metal substrate with diameter and length of 8 mm and 30 mm, respectively. The porous Cu interlayer was inserted in the middle of joining part. Two different porosities namely at 15 ppi—pores per inch (P15) and 25 ppi (P25) were used where the pore size was smaller at higher ppi. Configuration of soldering joint joined in this study is schematically shown in Fig. 1. 234 N of load was imposed at the upper jig during heating process in order to obtain uniform solder layer. The soldering process was conducted in the Japanese Industrial Standard, JIS Z 3191 furnace supplied with an Argon gas flow (Fig. 2). Three different temperatures of 267, 287 and 307 °C with a constant holding time of 300 s was set for the soldering parameters.

2.2 Isothermal Aging and Analysis Procedure

All soldered samples were placed in an oven for aging treatment tests at high temperature of 150 °C for duration of 100, 200 and 500 h. The temperature were controlled with accuracy of ±2 °C. Once the aging treatment completed, the aged samples were then analyzed for joint strength for tensile testing. The tensile test was carried out at room temperature using a Instron® Corporation Universal Testing (Model No. 3369, Norwood, MA, USA) at crosshead speed of 0.5 mm/min. The fracture surface of the solder joint after tensile test were observed and analyzed

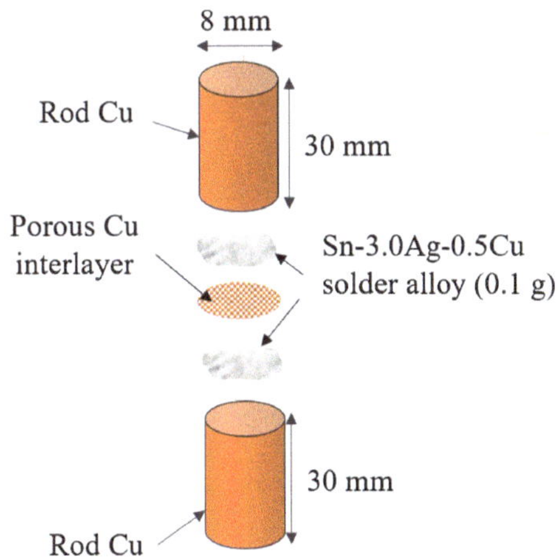

Fig. 1 Configuration of the solder joint

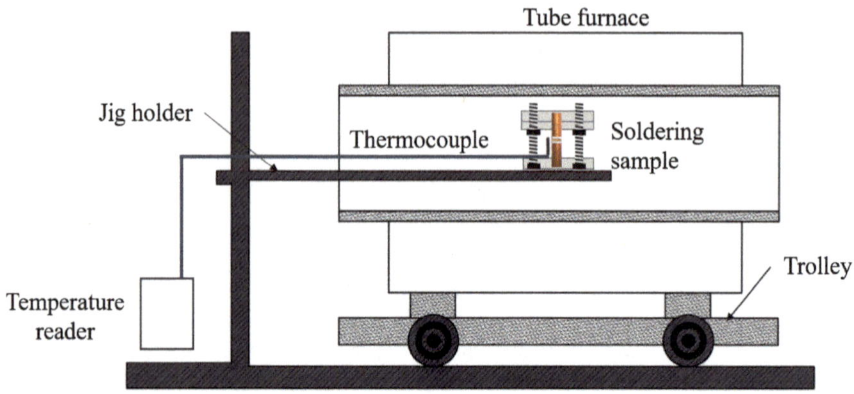

Fig. 2 Schematic of the furnace set-up for the soldering process

using a scanning electron microscope (SEM, Crest System (M) Sdn. Bhd., Eindhoven, Netherlands) equipped with energy-dispersive X-ray spectroscopy (SEM, Crest System (M) Sdn. Bhd., Eindhoven, Netherlands). The phase in fracture surface were analyzed by using X-ray diffraction (XRD, PaNalytical Empyrean, DKHS Holdings (Malaysia) Bhd., Almelo, Netherlands) which was operated in Cu-Kα radiation at 2θ diffraction angles ranging from 10° to 90°.

3 Results and Discussion

3.1 Effect of Porous Cu Interlayer on Joint Strength During Isothermal Aging Process

The average of pore diameter after being rolled for P15 and P25 porous Cu are 0.3 and 0.1 mm, respectively. These pore sizes influenced the penetration behavior of molten solder inside porous Cu during soldering process, thus affecting the solderability. Figure 3 presents the joint strength with respect to the isothermal aging and porous Cu addition in different soldering temperatures. It can be seen that the solder joint with P15 and P25 porous Cu interlayer show a significant reduction in the joint strength with increasing aging time. The strength for solder joint with P15 porous Cu at soldering temperature of 267 °C gradually decreased by 10 MPa for every 100 h of increasing aging time from 0 to 200 h and continued to decrease further by 10 MPa at 500 h. On the other hand sample with P25 porous Cu at the same soldering temperature exhibit the greater joint strength degradation of 25 MPa in the first 100 h which subsequently decreased by 5 MPa in the next 100 h increase in aging time and continue to decrease another 5 MPa at 500 h. Results further showed soldered sample of P15 and P25 porous Cu soldered at soldering time of 287 °C have roughly a similar pattern of joint strength degradation. Both

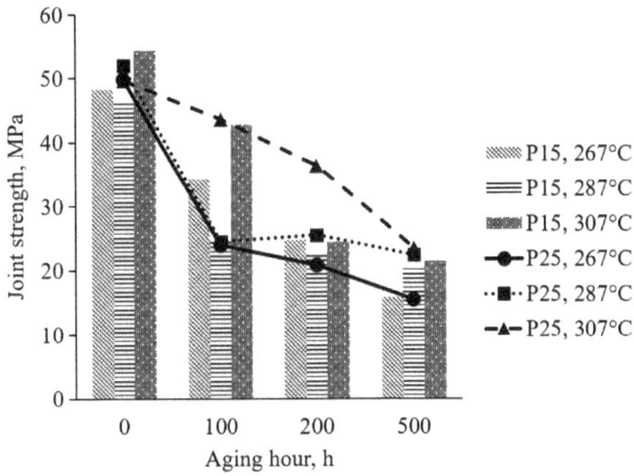

Fig. 3 Effect of porous Cu (P15 and P25) interlayer and soldering temperature on joint strength during isothermal aging

showed a sharp drop of more than 20 MPa during the first 100 h with P25 remains stable up to 500 h aging time while P15 decreases by 5 MPa in both the next 100 h and subsequent 300 h increase in aging time. Results also showed the solder joint with P15 and P25 porous Cu soldered at temperature of 307 °C have a better or similar joint strength than the rest of the solder joints. It was found that the strength did not cause any serious degradation up to 500 h. It can be concluded that soldering temperature of 307 °C shows better joining when compared to 267 and 287 °C with P25 having the edge over P15 as it has higher strength at 200 h and slightly better strength at 500 h aging time.

3.2 *Fracture Surface Analysis*

Fracture morphology was observed in order to understand the failure behavior of the aged solder joint added with porous Cu interlayer. The SEM photo-micrographs of fracture surface carried out after tensile test at 0, 100, 200 and 500 h for solder joints soldered with porous Cu interlayer of P15 and P25 are shown in Fig. 4a–h. Figure 4a, e showed many dimples appear in the typical fracture of as soldered samples (0 h aging time) for P15 and P25 solder joint, respectively. Fractured form of the porous Cu interlayer was obviously detected as shining dimples at the broken surface. As the aging time increased, the fracture surface appeared to have reduced in amount of dimples therefore has less roughness as compared to that of 0 h. This suggests brittle fractures occurred in the IMC layer which can be attributed to the breaking up of Cu_6Sn_5 of IMC layer that grows at excessive temperature condition. It can be explained by the large existence of Cu_6Sn pieces of the IMC phase at fracture surface resulting in lower joint strength especially at 500 h aging time ($\sim$20 MPa). This reduction in joint strength is caused by the coarsening of the grain which developed from the atomic reactions of the solder alloy and porous Cu during the isothermal aging. Therefore, it can be concluded that the penetration of

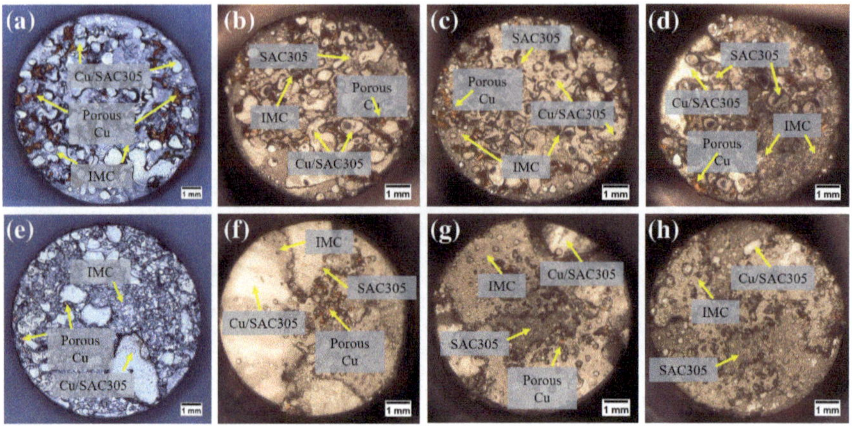

Fig. 4 Effect of isothermal aging at 150 °C for **a**, **e** 0 h, **b**, **f** 100 h, **c**, **g** 200 h and **d**, **h** 500 h on fracture morphology of solder joint added with porous Cu interlayer, **a–d** P15 and **e–h** P25 porous Cu interlayer

molten solder into porous Cu in solder joint tend to increase with increasing porosity and it significantly influenced fracture morphology and strength of the joint.

The XRD analysis was performed so as to examine the phase structure of IMC. The predominant peaks for all aging times are Cu and Sn element as shown in Fig. 5i–iv. However, the Cu_6Sn_5 IMC phase later becomes the predominant elements as the aging time increased. This may be due to the fact that more diffusion process of Cu-Sn atom occurred at porous Cu/SAC305 interface during penetration of molten solder during soldering process. Subsequently, further growth of IMC layer occurred during thermal storage thus influence the peak of IMC element at XRD results.

The percentage of fracture region are presented in Fig. 6 in order to show the relationship between the failure region and the aging time. It is observed that the fracture occurred at following regions; (i) Cu/SAC305 interface, (ii) porous Cu, (iii) inside solder and (iv) IMC phase of solder joint. The same findings were as reported in previous study [14]. It is found that mixed failure for fractures at the SAC305/Cu interface and inside solder were dominant for solder joint with P15 porous Cu. With increase in time of aging, the percentage of brittle failure (fractured at IMC) increased. In addition, bright dimples of broken porous Cu disintegrate. In contrast, the major fracture mode for solder joint with P25 porous Cu comprised ductile and the mixed mode failure at Cu/SAC305 interface and Cu_6Sn_5 of IMC. The growth of IMC layer and the roughness of the fracture surface influenced the mechanical characteristic of the solders. The increase of IMC thickness reduce the stress concentration at solder/porous Cu interface thus degrades the joint strength.

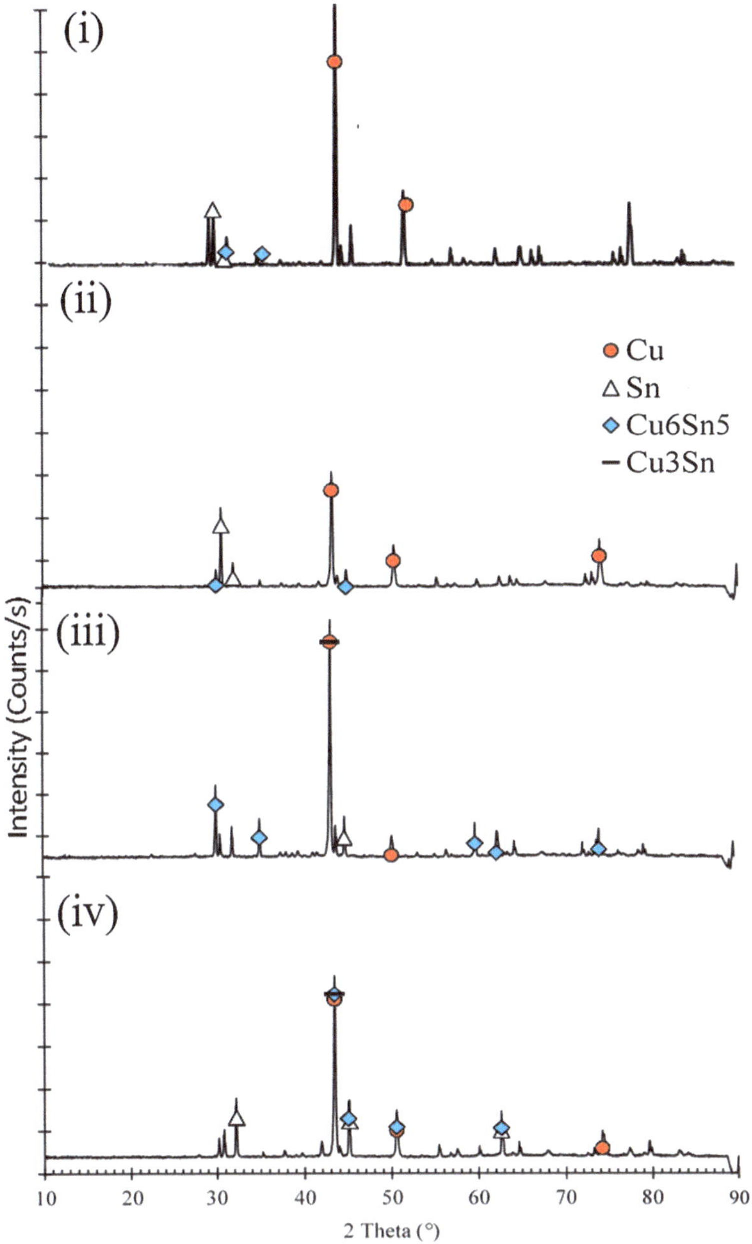

Fig. 5 XRD pattern for fracture surface of solder joint added with P15 porous Cu for (*i*) 0 h, (*ii*) 100 h, (*iii*) 200 h and (*iv*) 500 h aging time

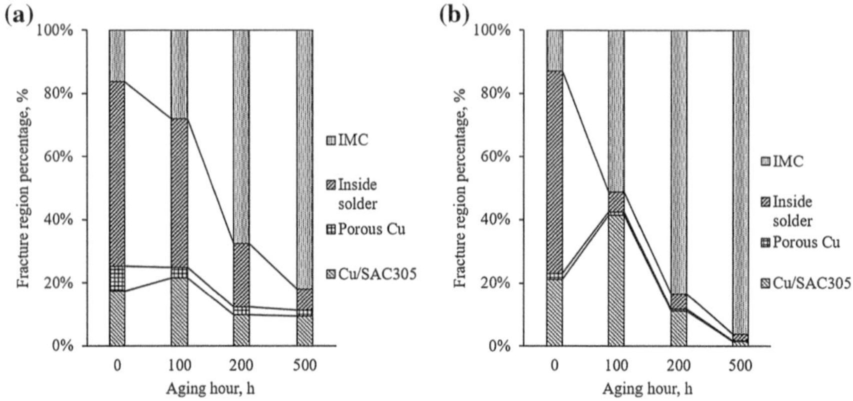

Fig. 6 Fracture mode percentage for solder joint added with **a** P15 and **b** P25 porous Cu against aging time

4 Conclusions

Present studies involve the effect of thermal aging on SAC305 solder joint with addition of porous Cu interlayer. Efforts have been focused on the assessment of tensile strength, microstructure analysis in the interfacial regions and fracture behavior. It can be concluded that the joint strength of the SAC305 added with porous Cu was influenced by soldering temperature, aging time and porosity of Cu. The joint strength of solder joint added with porous Cu decreased with increasing thermal aging time for the different soldering temperatures. Observations on the fracture morphology reveal the failure behavior occur at Cu/SAC305 interface, inside solder alloy and inside porous Cu. The IMC layer which consists of Cu_6Sn_5 compound was observed at the interfaces of Cu substrate and SAC305 solder alloy as well as porous Cu layer and the solder alloy. The IMC layer increased with increase in isothermal aging. It appears this IMC could be the main factor for decreasing joint strength during the isothermal aging.

Acknowledgements The author greatly appreciated the financial support of this research by the Fundamental Research Grant Scheme (FRGS, project number FP062-2015A) from University of Malaya and contribution from Nagaoka University of Technology, Japan for their support on this study.

References

1. Kim S-J, Kim K-S, Kim S-S et al (2008) Characteristics of Zn-Al-Cu alloys for high temperature solder application. Mater Trans 49:1531–1536
2. Chidambaram V, Hattel J, Hald J (2011) High-temperature lead-free solder alternatives. Microelectron Eng 88:981–989

3. Pang JHL, Xu L, Shi XQ et al (2004) Intermetallic growth studies on Sn-Ag-Cu lead-free solder joints. J Electron Mater 33:1219–1226
4. Shnawah DA, Sabri MFM, Badruddin IA et al (2013) Effect of Ag content and the minor alloying element Fe on the mechanical properties and microstructural stability of Sn-Ag-Cu solder alloy under high-temperature annealing. J Electron Mater 42:470–484
5. Huang Y, Xiu Z, Wu G et al (2016) Improving shear strength of Sn-3.0Ag-0.5Cu/Cu joints and suppressing intermetallic compounds layer growth by adding graphene nanosheets. Mater Lett 169:262–264
6. Kanchanomai C, Miyashita Y, Mutoh Y (2002) Low-cycle fatigue behavior of Sn-Ag, Sn-Ag-Cu, and Sn-Ag-Cu-Bi lead-free solders. J Electron Mater 31:456–465
7. Sharif A, Lim JZ, Made RI et al (2013) Pb-Free glass paste: a metallization-free die-attachment solution for high-temperature application on ceramic substrates. J Electron Mater 42:2667–2676
8. Amalu EH, Ekere NN (2012) High temperature reliability of lead-free solder joints in a flip chip assembly. J Mater Process Technol 212:471–483
9. Yoon JW, Noh BI, Choi JH, Jung SB (2011) Effect of adding Ce on interfacial reactions between Sn-Ag solder and Cu. J Mater Sci Mater Electron 22:745–750
10. Lin HJ, Chuang TH (2010) Effects of Ce and Zn additions on the microstructure and mechanical properties of Sn-3Ag-0.5Cu solder joints. J Alloys Compd 500:167–174
11. Cheng F, Gao F, Nishikawa H, Takemoto T (2009) Interaction behavior between the additives and Sn in Sn-3.0Ag-0.5Cu-based solder alloys and the relevant joint solderability. J Alloys Compd 472:530–534
12. Liu W, Lee N-C, Bachorik P (2013) An innovative composite solder preform for TLP bonding—microstructure and properties of die attach joints. In: 2013 IEEE 15th Electronics Packaging Technology Conference (EPTC 2013), pp 635–640
13. Yusof F, Hamdi M, Ariga T, Fadzil M (2013) Microstructure analysis of Brazed Sapphire to Inconel® 600 Using Porous Interlayer. 92:332–335
14. Jamadon NH, Tan AW, Yusof F et al (2016) Utilization of a Porous Cu interlayer for the enhancement of Pb-Free Sn-3.0Ag-0.5Cu solder joint. Metals (Basel) 6:220

On Amorphous Phase Formation in Dissimilar Friction Stir Welding of Aluminum to Steel

S.A. Hussein, A.S.M. Tahir and M. Awang

Abstract Amorphous phase which associated the joining of aluminum-to-steel via friction stir welding had been investigated in this study. AA5083, AA6061 and zinc coated steel were welded together by single pass friction stir welding. Different welding parameters of rotational and welding speeds had been used. Heat cycle measurement and scanning electron microscopy were used to investigate the effect of these welding parameter levels on the amorphous phase and intermetallic compound formations. The results showed that using of 2000 rpm with 20 mm/min as rotational and welding speeds, respectively, produced a long heat cycle. Such long heat cycle could eliminate the amorphous phase and promote thick intermetallic compounds, especially at the AA5083 to steel interface. However, when welding speed was increased to be 40 mm/min, a shorter heat cycle time was produced, and the amorphous phase was detected. Very high welding speed (300 mm/min) resulted in very short heat cycle with no obvious amorphous phase and very small amount of intermetallic compound at the aluminum-to-steel interface. The formation process of the amorphous phase is schematically explained in this study.

S.A. Hussein
Technical Instructor Training Institute, Middle Technical University, Baghdad, Iraq

A.S.M. Tahir
Faculty of Mechanical Engineering, Universiti Teknikal Malaysia Melaka (UTeM), Melaka, Malaysia

M. Awang (✉)
Mechanical Engineering Department, Universiti Teknologi PETRONAS, 32610 Seri Iskandar, Perak, Malaysia
e-mail: mokhtar_awang@utp.edu.my

M. Awang (ed.), *2nd International Conference on Mechanical, Manufacturing and Process Plant Engineering*, Lecture Notes in Mechanical Engineering,
DOI 10.1007/978-981-10-4232-4_6

1 Introduction

1.1 Hybrid Structure

Design that aims to produce light weight transports applications is a never-ending desire. A structure made from multi-materials (Bimetal, Trimetal, and Tetrametal) provide the best compromise solution of application that achieves weight reduction, performance improvement, and efficient energy. Moreover, topological structure optimization of this hybrid structure can help the integrated load-supporting implementation. Therefore, piece-by-piece replacement in a high strength-low weight structure is important to provide different functions in one structure. The integrated structure shown in Fig. 1 is an example of these types of structures. However, such structure requires efficient joining method of these dissimilar materials.

It is well known that low weight property of aluminum is related to its specific mass (density); 2700 kg/m^3 while iron and copper are 7800 kg/m^3 and 8500 kg/m^3 respectively, which represent one-third that of steel. Therefore, replacing steel parts by Al Alloys in the established steel framework structure can help the aimed hybrid structure. Integrated structures of steel and aluminum are widely used to attain high strength to weight ratio in transport application industries (Fig. 1). Efficient joining method of aluminum-to-steel (Al-to-steel) is a critical issue in manufacturing such hybrid structure. The difficulty of joining Al-to-steel using fusion welding methods is related to the advanced differences in melting temperature, thermal properties, and cooling rate after welding of both metallic alloys. Therefore, the joint strength is affected by dissolution of precipitates, weld defects, different microstructure of

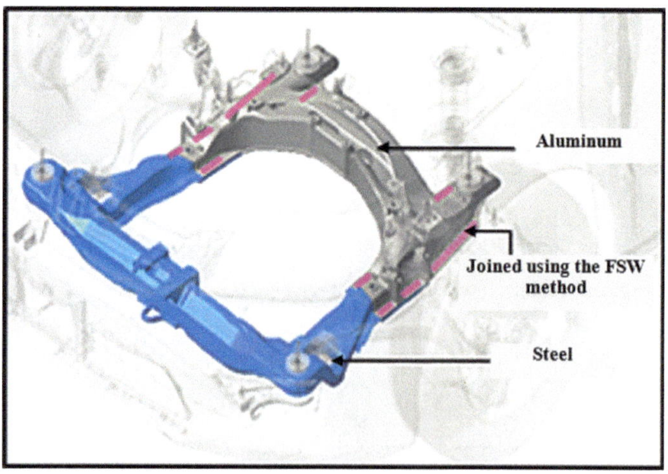

Fig. 1 Light weight product by Honda, dissimilar materials, and continuous FSW instead of bolt joints in new sub-frame of Honda Accord 2013 [1]

the diverse materials at the weld zone and promotion of thick intermetallic compound (IMC). Friction stir welding (FSW) can achieve efficient joint strength due to the solid state feature of this welding method [2].

1.2 Friction Stir Welding

FSW is an emerging mechanical solid-state welding method that is used to weld a wide range of materials (weldable/nonweldable) to produce hybrid structure [3]. It is a metalworking process in which the material does not melt and hence no resolidification process occurs [4].

During FSW process, the base material is severely deformed and heated by a rotating combined tool of shoulder and pin. The pin stirs the plasticized material, and the weld joint is formed by the softened metal [5]. The shoulder is the main heat generation source, and it also prevents the stirred material from leaking out. This part must be in contact with the workpieces when the tool makes a linear traverse along the abutting line. To avoid the movement that resulting from stir effect at the abutting line, the two workpieces should be clamped properly. Figure 2 shows the FSW requirements. Friction stir lap welding (FSLW) and friction stir butt welding (FSBW) are the main two configurations of this welding method. Recently, T-joint has been successfully made using this welding method by [6, 7].

Some of the main advantages of FSW have been summarized in Table 1.

1.3 Amorphous Phase

FSW is solid state welding method, but it also promotes IMC layer at the interface of Al-to-steel. Several researches have been conducted to characterize the IMC of the Al-to-steel joint made by FSW [2]. Springer et al. [9] stated that the pre-deformation of the parent materials caused by the FSW process will have pronounced effect on the composition and growth of the IMC layer, in which an amorphous phase is formed before the IMC layer formation. Ogura et al. [10] who noticed this amorphous phase stated that the amorphous phase is most likely a kneading stage that leads to the IMC formation when the applied heat is sufficient. Sun et al. [11] emphasized that the amorphous phase is caused mainly by mechanical alloying during the excessive plastic deformation, and not because of a thermally driven mechanism. They are convinced that the formation of IMCs is a thermally driven diffusion process. It is clear from the addressed literature that IMC is formed in the Al-to-steel joint made by FSW [12, 13]. An amorphous phase might also be formed before the IMC presence [10]. Very few studies discussed this phase, which implies more investigations that can explain the influential parameter on such phase.

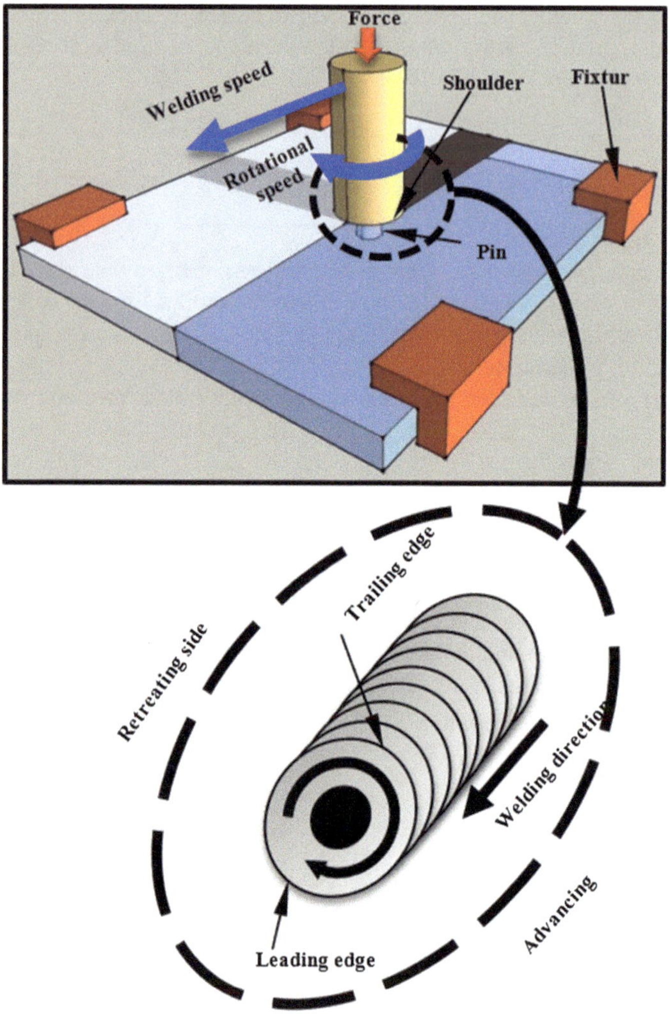

Fig. 2 Schematic of friction stir welding requirements

Table 1 The Main Advantages of FSW/FSP [3, 8]

Field description	Advantages
Defects	Closure of porosity, no crack formation, low dissolution of precipitates, low residual stresses, and low distortion
Fatigue and mechanical properties	Refinement of matrix grains, high tensile strength and fatigue, good elongation, can weld the non weldable materials/alloys with good joint efficiency
Environment, performance, and cost	No arc flash, no fumes, no spatter, energy efficient, versatile, joining the common joint-configurations (butt, lap, and T-joint), dissimilar material/alloy can be welded, no filler metals are needed, no shielding gas required in most cases, and low cost

In this study, a trial had been made to explain this amorphous phase and the affected welding process parameters. Besides, the joining of different types of aluminum alloys by single pass FSW could add further significant data to the outcomes of this study.

2 Experimental Works

The workpieces materials were 3 mm thick AA5083-H112, AA6061-T6 sheets and 1.5 mm thick zinc coated-steel sheet. The compositions of these metallic alloys per the manufacturer's data sheet are shown in Table 2. Vertical CNC milling machine (HASS VF-1D) was used. The arrangement of the plates is shown in Fig. 3. The welding tool made from tungsten carbide and tool steel-H13 is shown in Fig. 4. Tool shoulder was dipped into the workpieces by 0.3 mm to insure the contact incidents between the tool and the workpiece. The pin tip was not plunged into the steel, rather, diffusion technique was used [2]. The welding parameters are shown in Table 3. The rotational and downward speeds during the plunging stage were 1000 rpm and 5 mm/min, respectively, which could avoid incoming high forces [14]. Heat cycle was measured experimentally at the Al-to-steel interface by using a USB TC-08 thermocouple (range of 0–1820 °C) with 8-port capacity, as shown in Fig. 3. Scanning electron microscopy (SEM) was then used to investigate the amorphous phase and IMC formations at the weld interface microstructure. The microstructure samples were prepared for the microstructure tests by cold mounting to avoid associated incoming heat during the hot mounting process. The surfaces were grinding with abrasive paper grit 500 and 1200, and then polished with 9, 6, 3 and 1 μm diamond suspensions.

Table 2 Chemical and mechanical properties of the materials

Material	Chemical composition wt%	Actual UTS (MPa)
AA5083-H112	0.08 Si, 0.27 Fe, 0.03 Cu, 0.65 Mn, 4.71 Mg, 0.08 Cr, 0.04 Zn, 0.02 Ti, balance Al	313
AA6061-T6	0.67 Si, 0.32 Fe, 0.32 Cu, 0.014 Mn, 1.06 Mg, 0.21 Cr, 0.007 Zn, 0.02 Ti, other 0.05, balance Al	299
Zinc-coated-steel (JIS G3313)	0.0204 C, 0.027 Si, 0.199 Mn, 0.0079 P, 0.0083 S. Zinc coated weight is 19 g/m^2 on both surfaces	282

Note The ultimate tensile strength (UTS) of this table is the average of four tested samples for each tabulated material

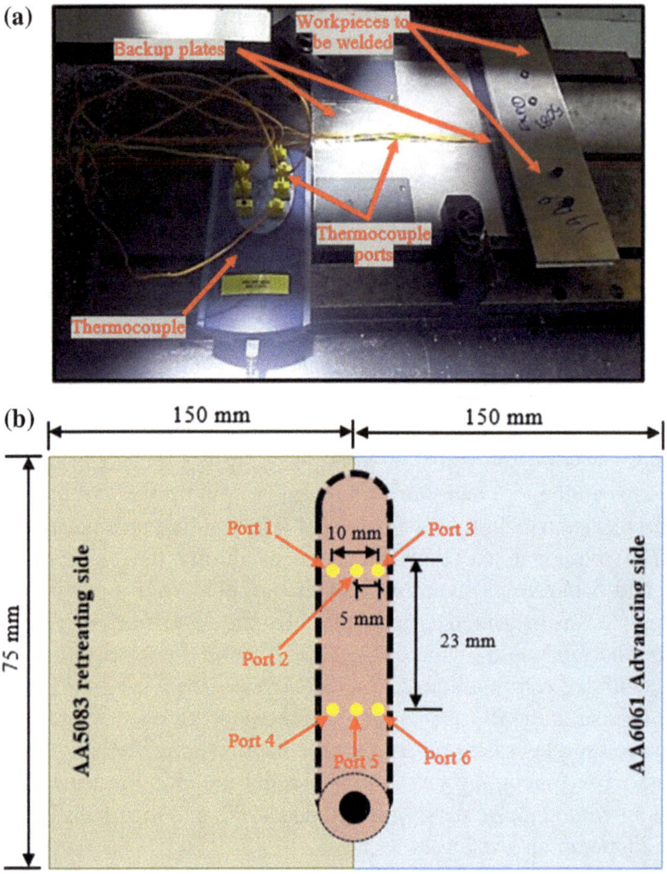

Fig. 3 Experimental setup **a** the workpieces fixture and the CNC table **b** schematic diagram of the thermocouple ports positions at the Al-to-steel interface

3 Results and Discussion

Rotational speed (N) and welding speed (v) affected the heat cycle (thermal history) of the weld. Peak temperature (Tp) of welding process when N = 1800 rpm (Fig. 5) was lower than the Tp when N = 2000 rpm (Fig. 6). It is well known that increasing N factor will lead to an increase in the heat input [15]. Contrarily, Figs. 5 and 6 show that Tp decreased with an increase in the welding speed, i.e., 20 mm/min provided higher weld temperature as compared to 300 mm/min. A large difference in welding speed tends to reduce the heat input to the workpieces during the FSW process [16]. This phenomenon is related to the long frictional time, and high plastic deformation occurred along the welding line [17]. This welding speed significantly affected the time history of the heat cycles (Figs. 5 and 6) as well. In other words, low welding

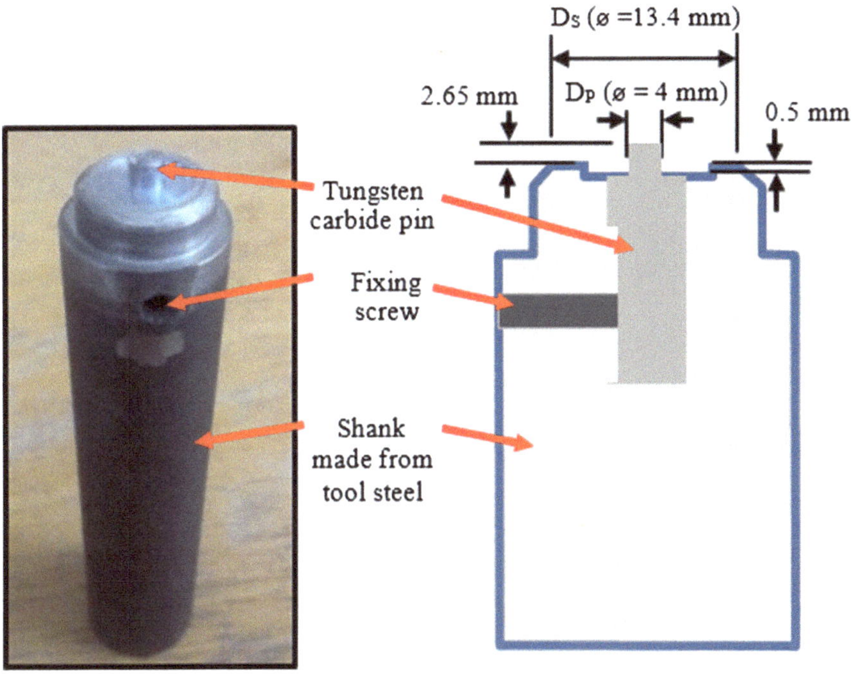

Fig. 4 Design specification of the FSW tool

Table 3 The welding parameters

Rotational speed (rpm)	Welding speeds (mm/min)
1800	20, 300
2000	20, 40, 300

speed (such as 20 mm/min) showed a longer time for the workpiece to be under hot condition.

Due to the differences in materials flow, the temperature distribution is asymmetric as claimed by Jacquin et al. [17]. The temperature distributions at the advancing side (AS) and retreating side (RS) are explained in Figs. 5 and 6). Nandan et al. [18] and Rao et al. [5] stated that pin rotation and welding direction would be similar on the AS and opposite on the RS of the weld. Therefore, high heat is concentrated at the high flow of metal region (AS) because of high plastic deformation [5, 18]. In the present study, the hotter zone was at AS when the welding speed was 20 mm/min. However, this was not occurred when the welding speed was very high (300 mm/min). It is most likely that this temperature distribution was affected by the rate of welding speed and weld pitch (rev/mm). The high weld pitch might not provide enough time to stabilize the temperature, especially, with the dissimilarities in the properties of the welded Al Alloys, such as thermal conductivity which affected by the temperature rise.

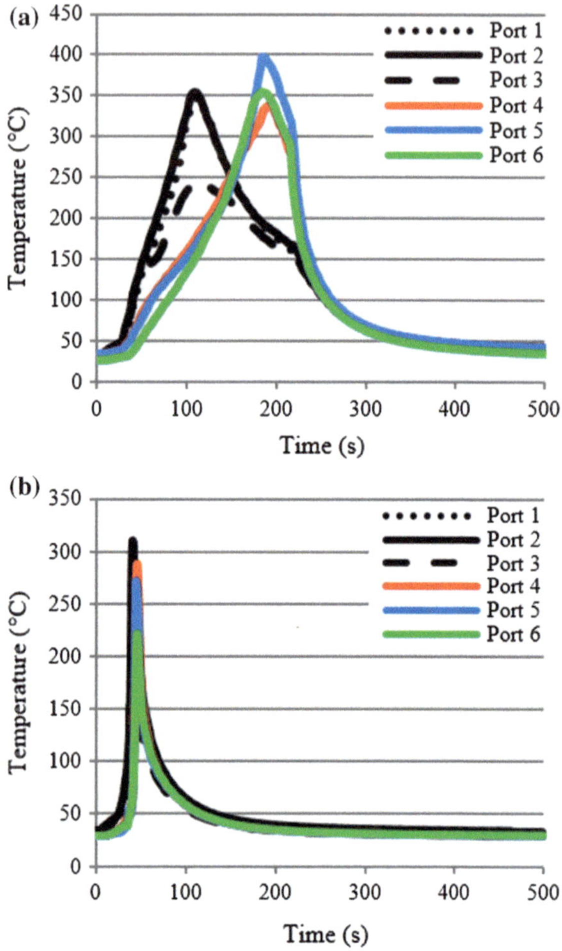

Fig. 5 Heat cycle of weld made by 1800 rpm with **a** 20 mm/min, **b** 300 mm/min

Consequently, heat amount and tool duration at each point along the weld line were very important factors in this heat distribution issue. The heat cycle of 2000 rpm with 40 mm/min shown in Fig. 7 resulted in shorter heat cycle compare to 2000 rpm with 20 mm/min, but longer than 2000 rpm with 300 mm/min.

During the microstructure investigation, the weld made by using high welding speed 300 mm/min did not show a noticeable amorphous phase, and very small discontinuous IMC layer could be observed, as shown in Fig. 8. Besides, no obvious amorphous phase could be noticed at the Al-to-steel interface when 20 mm/min was used and for both rotational speeds, as shown in Fig. 9. IMC layer was formed at both sides with different thicknesses. In general, these IMCs were thick at the AA5083 side in which showed some cracks, while AA6061 side promoted thin free of cracks layer for the case of v = 20 mm/min. Furthermore, it was very difficult to find when v = 300 mm/min.

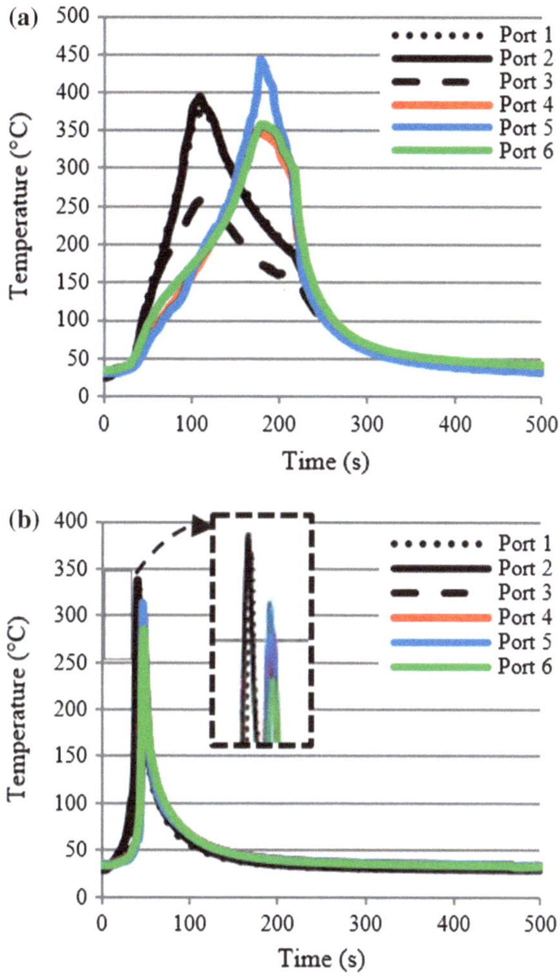

Fig. 6 Heat cycle of weld made by 2000 rpm with **a** 20 mm/min, **b** 300 mm/min

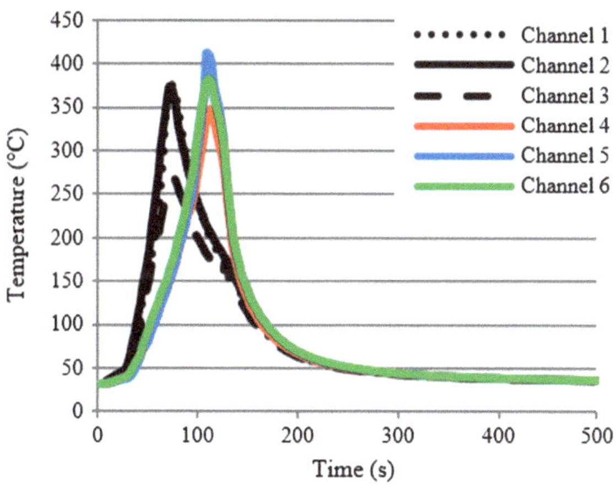

Fig. 7 Heat cycle of weld made by 2000 rpm with 40 mm/min

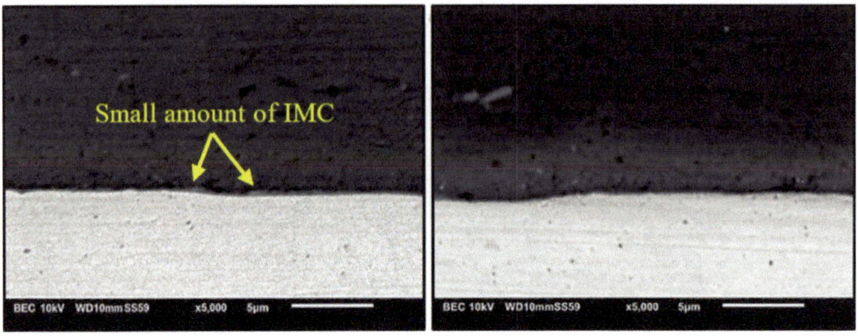

Fig. 8 Very small amount of the IMC at the AA5083-to-steel side for weld made by using v = 300 mm/min

The peak temperature of port 5 and the cooling time of the heat cycle measurement were lower when v = 40 mm/min was used as compared to 20 mm/min. This difference in the heat cycle promoted amorphous phase and thinner IMC layer thickness was formed. The temperature at the AA6061 side was higher than the temperature at the AA5083 for the rotational speed of 2000 rpm with 20 and 40 mm/min as welding speeds. This was significant observation as AA5083 promoted thick IMC than the AA6061 though the lower temperature at that side. This phenomenon can be related to the Si presence at the AA6061 plate. Kobayashi and Yakou [19] believed that IMC layer is inhibited by the existence of Si atoms which are higher in AA6061 material as compared to the AA5083 one (Table 2). Cracks were found at the Al-to-steel interface (Fig. 9c), which denoted a deleterious effect on the joint strength [2]. Hot and long heat cycle led to thick IMC when N = 2000 rpm and v = 20 mm/min were used.

In Fig. 9 (e and f), an amorphous phase is clearly shown. The increase of welding speed leads to lower plastic deformation [20]. Therefore, 40 mm/min led to lower plastic deformation than that obtained by using 20 mm/min. On the other side, increasing the rotational speed will increase the plastic deformation. It is believed that increasing the heat cycle could help the transformation of this amorphous phase to the IMC phase, which would further increase the layer thickness [11]. When welding speed was set to 20 mm/min, no obvious amorphous phase was detected because of the long heat cycle. It is worthy to mention that excessive plastic deformation could be achieved by considering other parameters such as threaded pin profile or big tool size [10]. Therefore, high plastic deformation together with low heat cycle condition (2000 rpm, 40 mm/min), were likely to be the main cause of this phase existence. When plastic deformation and heat cycle were high (2000 rpm, 20 mm/min) this phase was not observed. Setting N = 2000 rpm with v = 300 mm/min produced lower heat cycle and plastic deformation, and thereby no amorphous phase was detected though very small IMC was observed. The AS was subjected to more plastic deformation, which induced more amorphous phase amount on this side (Fig. 9d), as explained and summarized

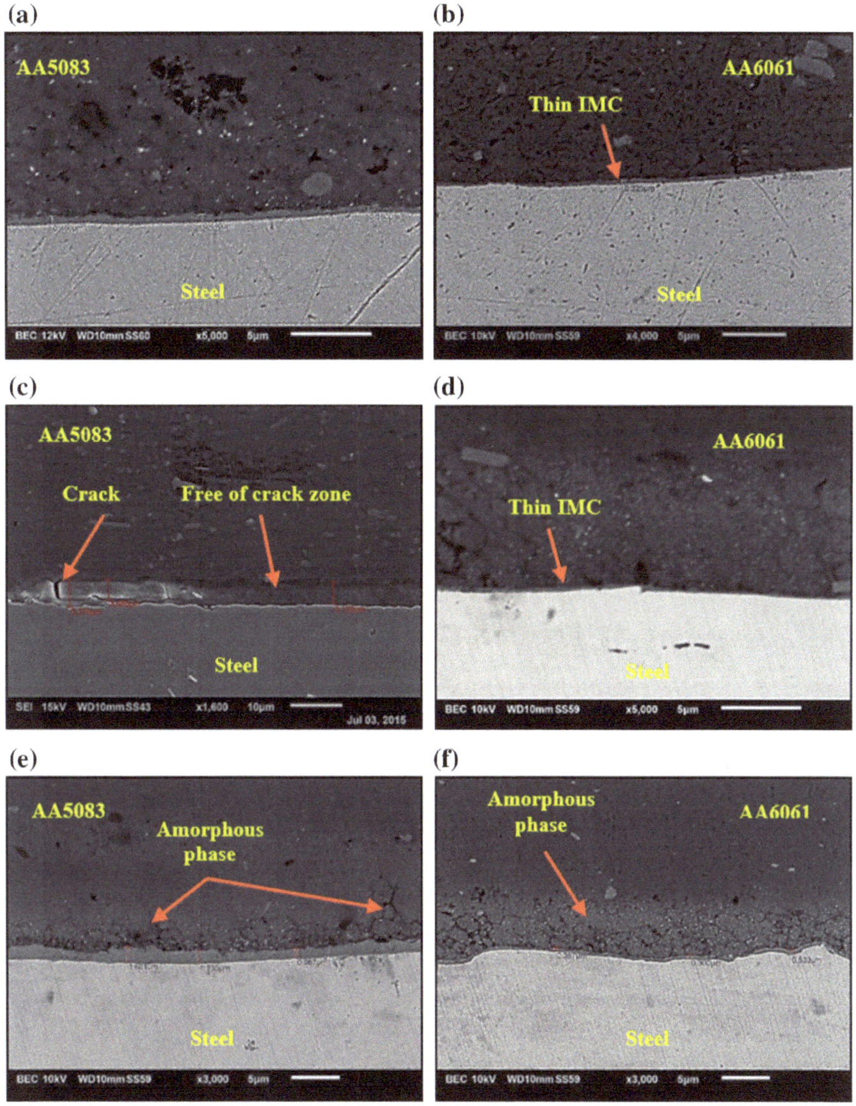

Fig. 9 Al-to-steel interfaces for different welding parameters at both dissimilar Al Alloys **a** $N = 1800$ rpm with $v = 20$ mm/min AA5083-to-steel, **b** $N = 1800$ rpm with $v = 20$ mm/min AA6061-to-steel side, **c** $N = 1800$ rpm with $v = 20$ mm/min AA5083-to-steel, **d** $N = 2000$ rpm with $v = 20$ mm/min AA6061-to-steel side, **e** $N = 2000$ rpm with $v = 40$ mm/min AA5083-to-steel, and **f** $N = 2000$ rpm with $v = 40$ mm/min AA6061-to-steel side

in Fig. 10. It is worthy to mention that transmission electron microscopy (TEM) technique is better suited for microstructural investigation due to the much higher magnification and resolution abilities [21]. This issue can be assumed as a limitation in this study.

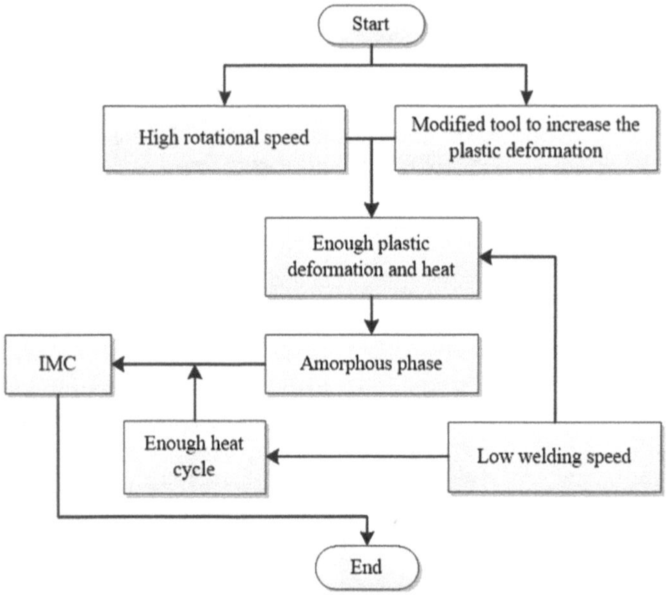

Fig. 10 Amorphous phase formation mechanism and its transformation process

4 Conclusion

Heat cycle and microstructure have been investigated in this study to evaluate the amorphous phase formation of Al-to- steel joint made by FSW. Remarkable points can be summarized as follows:

1. Amorphous phase is affected by the amount of heat and plastic deformation which resulted from using different welding process parameters. High plastic deformation together with short heat cycle led to this phase as explained by the case of 2000 rpm with 40 mm/min in this study. Though there was high plastic deformation, welding parameters that offer long heat cycle could eliminate the presence of the amorphous phase. For example, using 2000 rpm with 20 mm/min as rotational and welding speeds, respectively, had transformed that amorphous phase into IMCs phase. Therefore, the amorphous phase was not observed with the using of these welding process parameters. Very high welding speed (300 mm/min) also cannot produce the amorphous phase at the Al-to-steel interface due to the low associated plastic deformation.
2. The types of Al Alloys can also affect the formation process of amorphous phase and the related IMC. At the AA6061 (advancing side), higher amorphous phase is believed to be caused by the higher plastic deformation at that side. Besides, thinner IMC layer was detected at that side for different welding parameters, which can be related to the silicone affinity in this type of Al Alloy.

References

1. Kusuda Y (2013) Honda develops robotized FSW technology to weld steel and aluminum and applied it to a mass-production vehicle. Ind Robot An Int J 40:208–212
2. Hussein SA, Tahir ASM, Hadzley AB (2015) Characteristics of aluminum-to-steel joint made by friction stir welding: a review. Mater Today Commun 5:32–49
3. Mishra RS, Ma ZY (2005) Friction stir welding and processing. Mater Sci Eng R Reports 50:1–78
4. Koilraj M, Sundareswaran V, Vijayan S et al (2012) Friction stir welding of dissimilar aluminum alloys AA2219 to AA5083—Optimization of process parameters using Taguchi technique. Mater Des 42:1–7
5. Rao D, Huber K, Heerens J et al (2013) Asymmetric mechanical properties and tensile behaviour prediction of aluminium alloy 5083 friction stir welding joints. Mater Sci Eng A 565:44–50
6. Zhao Y, Zhou L, Wang Q et al (2014) Defects and tensile properties of 6013 aluminum alloy T-joints by friction stir welding. Mater Des 57:146–155
7. Hou X, Yang X, Cui L et al (2014) Influences of joint geometry on defects and mechanical properties of friction stir welded AA6061-T4 T-joints. Mater Des 53:106–117
8. Gibson BT, Lammlein DH, Prater TJ et al (2014) Friction stir welding: process, automation, and control. J Manuf Process 16:56–73
9. Springer H, Kostka A, Dos Santos JF et al (2011) Influence of intermetallic phases and Kirkendall-porosity on the mechanical properties of joints between steel and aluminium alloys. Mater Sci Eng A 528:4630–4642
10. Ogura T, Saito Y, Nishida T et al (2012) Partitioning evaluation of mechanical properties and the interfacial microstructure in a friction stir welded aluminum alloy/stainless steel lap joint. Scr Mater 66:531–534
11. Sun YF, Fujii H, Takaki N et al (2013) Microstructure and mechanical properties of dissimilar Al alloy/steel joints prepared by a flat spot friction stir welding technique. Mater Des 47:350–357
12. Ramachandran KK, Murugan N, Shashi Kumar S (2015) Effect of tool axis offset and geometry of tool pin profile on the characteristics of friction stir welded dissimilar joints of aluminum alloy AA5052 and HSLA steel. Mater Sci Eng A 639:219–233
13. Fei X, Jin X, Ye Y et al (2016) Effect of pre-hole offset on the property of the joint during laser-assisted friction stir welding of dissimilar metals steel and aluminum alloys. Mater Sci Eng A 653:43–52
14. Hussein SA, Tahir ASM, Izamshah R (2015) Generated forces and heat during the critical stages of friction stir welding and processing. J Mech Sci Technol 29:4319–4328
15. Frigaard Ø, Grong Ø, Midling OT (2001) A process model for friction stir welding of age hardening aluminum alloys. Metall Mater Trans A 32:1189–1200
16. Cam G, Mistikoglu S (2014) Recent developments in friction stir welding of al-Alloys. J Mater Eng Perform 23:1936–1953
17. Jacquin D, De Meester B, Simar A et al (2011) A simple Eulerian thermomechanical modeling of friction stir welding. J Mater Process Technol 211:57–65
18. Nandan R, Roy GG, Lienert TJ et al (2007) Three-dimensional heat and material flow during friction stir welding of mild steel. Acta Mater 55:883–895
19. Kobayashi S, Yakou T (2002) Control of intermetallic compound layers at interface between steel and aluminum by diffusion-treatment. Mater Sci Eng A 338:44–53
20. Liu HJ, Hou JC, Guo H (2013) Effect of welding speed on microstructure and mechanical properties of self-reacting friction stir welded 6061-T6 aluminum alloy. Mater Des 50: 872–878
21. Girard M, Huneau B, Genevois C et al (2010) Friction stir diffusion bonding of dissimilar metals. Sci Technol Weld Join 15:661–665

Influence of Loading Rate on Deformation Behaviour and Sealing Performance of Spiral Wound Gasket in Flange Joint

N. Rino Nelson, N. Siva Prasad and A.S. Sekhar

Abstract Owing to increase in the demand for power, a number of power plants have been installed in the recent past. One of the vital components in these plants is gasketed flange joint. The behaviour of gasketed flange joint is highly dependent on the deformation characteristics of gasket material. In the present study, the deformation characteristic of spiral wound gasket along thickness direction is determined experimentally for different loading rates. Gasket has high nonlinearity under both loading and unloading, with hysteresis. Based on its characteristics, the flange joint is analysed to study its performance and leakage behaviour under static condition. 3D finite element model of flange joint is developed by considering the gasket as interface entity. Finite element analysis of flange joint is performed by including the nonlinear hysteretic behaviour of gasket, under different loading rate and frictional contact between joint members. The influence of different loading and unloading rates is emphasized on the sealing performance of flange joint using leakage pressure. The gasket deforms more when loaded at low rate. This phenomenon also affects the ability to withstand internal fluid without leakage. The maximum safe pressure without leakage increases, when gasket is loaded and unloaded at low rate.

Nomenclature

F	Total bolt preload
H_D and H_T	Hydrostatic end forces
m	Gasket maintenance factor
P	Internal fluid pressure

N.R. Nelson (✉) · A.S. Sekhar
Mechanical Engineering, Indian Institute of Technology Madras, Chennai 600036, India
e-mail: nrinonelson@gmail.com

A.S. Sekhar
e-mail: as_sekhar@iitm.ac.in

N.S. Prasad
GITAM University, Hyderabad 502329, India
e-mail: sivacae@yahoo.co.in

M. Awang (ed.), *2nd International Conference on Mechanical, Manufacturing and Process Plant Engineering*, Lecture Notes in Mechanical Engineering,
DOI 10.1007/978-981-10-4232-4_7

ANSI	American National Standards Institute
FEA	Finite element analysis
RSL	Residual stress line
RVE	Representative volume element
SWG	Spiral wound gasket
SS	Stainless steel

1 Introduction

In recent years, many power plants have been installed, due to high demand for power generation. Flange joints are widely used in power plants for connecting long pipelines, head and shell portion in heat exchangers, nozzles and man holes. Gasket is primarily provided in the flange interface, to overcome surface imperfections. For high pressure and temperature operation, spiral wound gaskets (SWG) with soft fillers are preferred due to its stability. It is a semi-metallic gasket with stainless steel ring and flexible graphite ring wounded under radial compression. A section of spiral wound gasket is shown in Fig. 1, with inner and outer rings. The deformation characteristic of sealing ring is nonlinear, due to both material behaviour and geometric configuration.

Mathan and Siva Prasad [1] evaluated the properties of SWG considering representative volume element (RVE) by homogenisation method. RVE is subjected to various loading conditions to determine its averaged elastic and Hill's model responses. Nelson et al. [2] characterized the deformation behaviour of gasket by incorporating the individual component behaviour in SWG, treating them as separate entity. A six degree polynomial equation is used by Fukuoka and Takaki [3] to characterize both loading and unloading gasket behaviour. The influence of internal pressure in gasket load drop is analytically proposed by Nagy [4], using piecewise linear model for gasket characterization. The load drop in gasket is determined by the loosening coefficient, which is a function of gasket material and geometry. Nassar et al. [5] investigated the effect of loading history in gasket and bolt tightening patterns. When star or sequential tightening pattern is followed, all the other bolts get loosened other than the tightening bolt, resulting in local loading and

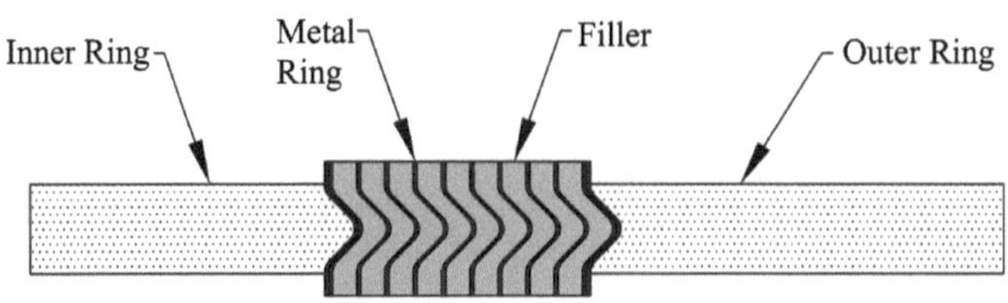

Fig. 1 A section of spiral wound gasket

unloading of gasket. The scatter in bolt load is observed to be more in flange joint with SWG. This scatter gets reduced on applying multiple tightening pass for above said patterns. The compression of gasket during assembly stage is a key factor for safe operation of flange joint, which in turn depends on its stiffness. In addition to stiffness, joint's leakage resistance also depends on the clamping force, joint rigidity, resistance to creep and surface texture of mating faces [6].

The primary failure in flange joint is leakage of internal fluid. The behaviour of flange joint is studied by Estrada [7] using contact finite element analysis (FEA). The accurate evaluation of gasket stress is very important to study the interaction between joint members. Bouzid and Galai [8] proposed an analytical model to predict the joint tightness along with flange rotation and relaxation in bolt load, based on flexibility of the joint and member interaction. In many cases, the gasket member alone fails in flange joint. Jenco and Hunt [9] revealed the common issues of gasket failure due to insufficient compression, gasket buckling and crushing. The limiting size for clearance between gasket and groove is provided by Roos et al. [10], which ranges between 1.0 and 1.5 mm. In general, gasket is observed to fail due to inward buckling/collapse, when it is provided without support rings. The buckling of gasket results in unwinding of sealing rings due to turbulence of internal fluid [11]. During compression, the distribution of stress becomes non-uniform in buckled gaskets, which is highly prone to leakage on pressurization. Nassar and Alkelani [12] investigated the effect of tightening speed on stress distribution in joint. The clamp load in soft gasket reduces with increase in tightening speed and vice versa in hard gasket.

SWG is observed to have high nonlinearity and hysteresis during loading and unloading [2]. It also has considerable amount of permanent deformation in gasket even after unloading the gasket. This viscos behaviour of gasket is very important for material characterization. In literature, this effect due to varying loading rate is not investigated, which might have higher influence on performance of flange joint.

In the present study, deformation characteristics of SWG under different loading rates are determined. Uniaxial compression testing is performed to determine the characteristics for 200, 500 and 1000 N/s loading rate. A 3D finite element model of flange joint is developed considering these deformation characteristics and contact between joint members. The effect of loading rate on sealing behaviour of flange joint is also analyzed for different bolt preload under internal fluid pressure.

2 Experimental Procedure

The behaviour of flange joint depends on deformation characteristics of gasket material. Spiral wound gasket is a semi metallic gasket with stainless steel (SS) windings and graphite filler. The behaviour of gasket along thickness direction is more important compared to other two directions. The deformation characteristics of this gasket are determined from uniaxial compression testing. The test set-up used for the present study is shown in Fig. 2. Here, vertical universal testing

Fig. 2 Experimental set up for uniaxial compression of gasket specimen

machine is used to apply the axial compressive load on specimen. This machine is self-built with extensometer and other sensors to measure the deformation in specimen, during application and removal of load. It also has the ability to control loading and unloading rate of specimen. The entire machine is controlled using a control system, through which the rate of loading and maximum compressive load is pre-set. The data acquisition system is used to retrieve the output data from experiment, both stress and deformation in the specimen.

The gasket specimen is subjected to gradual compressive load from 0 to 15 tonnes, based on specific loading rate. After loading the gasket to 15 tonnes, the load is reduced gradually to zero, defining the unloading characteristics of material. The initial (before loading) and final (after unloading) specimen thicknesses are measured, to determine the permanent deformation (plastic deformation) in gasket on application of load. The same procedure is repeated for 30 tonnes, in order to obtain another unloading curve. The loading and unloading of specimen is carried out at controlled load rate viz., 200, 500 and 1000 N/s. The specimen is kept in the loaded position for 10 min at the maximum load (15 or 30 tonnes), before relaxing the load. It is done on specimen to reach stabilization. Both loading and unloading is carried out at the same controlled rate.

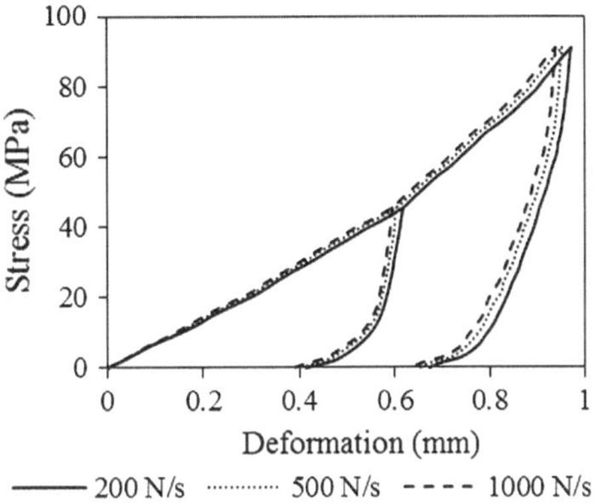

Fig. 3 Stress displacement relation of SWG for different loading rates

3 Gasket Characteristics at Different Loading Rate

Generally, the gasket materials used in flange joint are comparatively softer than other joint members. They possess high nonlinear characteristics and strong hysteresis. The leakage behaviour of flange joint is highly dependent on this characteristic i.e., deformation along thickness direction. The experiments are conducted on gasket specimen, by applying compressive load at different loading and unloading rates. The gasket material used is semi-metallic spiral wound gasket, with stainless steel (metal) rings and soft graphite (non-metallic) filler material. Usually, spiral wound gaskets are provided with inner and outer supporting rings. The sealing ring is composed of radially wounded SS and graphite rings, which helps to prevent leakage. The deformation of gasket is concentrated on sealing ring because, the inner and outer supporting rings are provided for stability purpose to prevent over compression. The thickness of gasket sealing ring is 4.5 mm; whereas supporting ring is 3 mm. So, on application of compressive load, the sealing ring alone gets compressed first. The deformation behaviour of gasket for different loading rates is shown in Fig. 3, against the corresponding stress values. The gasket material shows strong hysteresis and viscoplastic behaviour. On loading, gasket is subjected to both elastic and plastic deformation; whereas on unloading, only elastic deformation is recovered. The gasket deformation increases with decrease in loading rate. This might be due to increase in material flowability, when loaded at a low rate.

4 Flange Joint with Gasket

In this study, ANSI B16.5 Class 600, 80 mm NPS flange with weld neck and raised face is considered. Flange and bolt materials are carbon steel and high chromium steel respectively. The properties used to characterize the flange joint members are shown in Table 1. Except gasket, the other joint members are characterized using linear elastic material model to include only their elastic material behaviour.

Table 1 Material properties of joint members

Member	Young's modulus (GPa)	Poisson's ratio
Flange	195	0.3
Bolt	203	0.3

4.1 FE Model of Flange Joint

Finite element model of flange joint is developed and analyzed using Ansys v14.5. The flange geometry is discretized using SOLID185, 3D 8-noded element with 3 DOFs at each node [13]; whereas bolt by SOLID90. These solid elements have translation along three perpendicular directions as its DOF. The bolt and flange members are modelled as separate entity and contact is established between their interfaces. The bolt-nut connector is modelled as a single entity without considering threads. For FE modelling, the hexagonal bolt head and nut are modelled as cylindrical entity.

The contact behaviour is modelled using CONTA174 and TARGE170. Surface to surface contact with 0.2 friction coefficient is established in flange-bolt interface. Here, comparatively stronger member, flange is taken as target and bolt as contact region. The primary component in gasketed flange joint, gasket is modelled and discretized using interface elements, INTER195. It has a unique characteristic of having one element along thickness direction between the mating surfaces, which controls its deformation behaviour. Pretension applied to the bolt during assembly

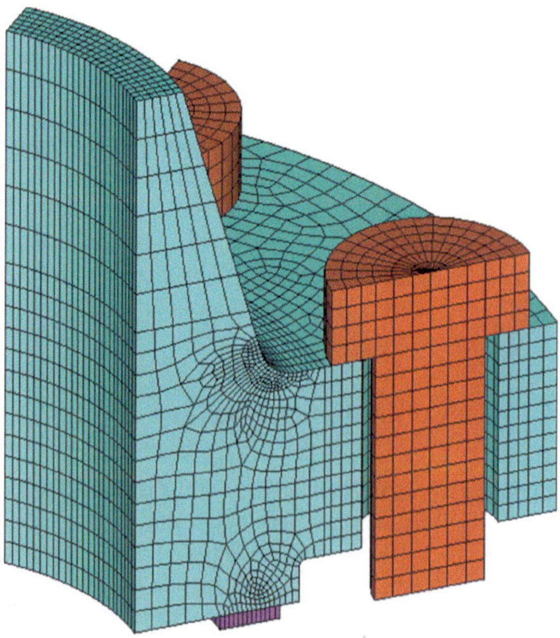

Fig. 4 FE model of flange joint

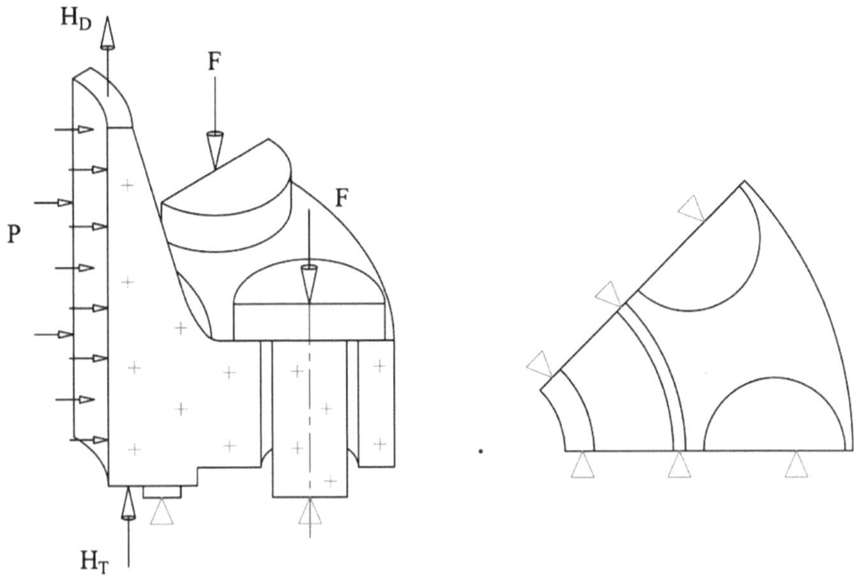

Fig. 5 Boundary condition

stage, for holding flanges together is simulated using a special element, PRETS179. A node on the specified plane is taken as pretension node and all other nodes on same plane of the entity are connected to it. The pretension load is directly applied on to the pretention node. Figure 4 shows the meshed FE model of the flange joint.

Flange joint is a cyclic symmetric component, due to which 45° segment of it is considered, covering two half bolt region and ligament region between them. One of the flanges and gasket with half thickness are modelled, considering the longitudinal symmetry. The bolt preload (F) is applied on each bolt during assembly stage and internal fluid pressure (P) is applied on pressurization stage. The hydrostatic end forces H_D and H_T [10], due to internal fluid pressure are also considered. The circumferential displacement is constrained on 0° and 45° surfaces of flange joint, in order to specify the cyclic symmetric condition. The bottom surface of the gasket is restrained along longitudinal direction (see in Fig. 5), to specify the longitudinal symmetry.

5 Gasket Behaviour

The deformation behaviour of sealing ring alone decides the sealing performance of flange joint. The flexible graphite filler used in gasket helps in sealing the mating surfaces, while SS ring provides stability to the sealing ring. The SS rings preformed in V-grooved shape, to provide the spring back effect in gasket during unloading. These V-grooves are placed in such a way that the opened end of ‘V’ is along the inner circumference. During assembly stage, the compressive load makes the graphite filler to squeeze into the irregularities in mating surface and SS ring also gets compressed. When internal pressure is applied, the flanges tend to move away from each other. This flange opening results in reduction of compressive load in gasket. When the flanges open, the compressed V-shaped metal (SS) rings produce partial spring back effect, along with graphite filler. The compressive stress remaining in sealing ring after spring back, is termed as residual stress in gasket. The leakage occurs, when the spring back effect of gasket sealing ring is not enough to retain the internal fluid at operating pressure.

5.1 Gasket Contact Stress Distribution

During assembly stage, the compressive stress in gasket follows the loading curve in Fig. 3; whereas during pressurization stage it depends on the unloading characteristics of the gasket material. The distribution of gasket stress along the radial width of the gasket varies from inner to outer diameter. This variation is due to the eccentric transfer of bolt load from bolt to gasket via flange member. The gasket stress is more along the outer circumference compared to inner edge, due to proximity towards bolt region.

On application of internal pressure, the gasket compression stress relaxes due to opening of flange joint. The internal fluid pressure tends to move the mating surfaces apart from each other. The hydrostatic end force acting on the area between flange and gasket inner diameter, results in more relaxation along the inner circumference of gasket compared to the radial width. The variation in gasket stress distribution along radial width of gasket at 2 and 10 MPa internal pressure and 120 kN total bolt preload for different loading rate is shown in Fig. 6. The variation in gasket stress along radial width increases with increase in loading rate and internal fluid pressure.

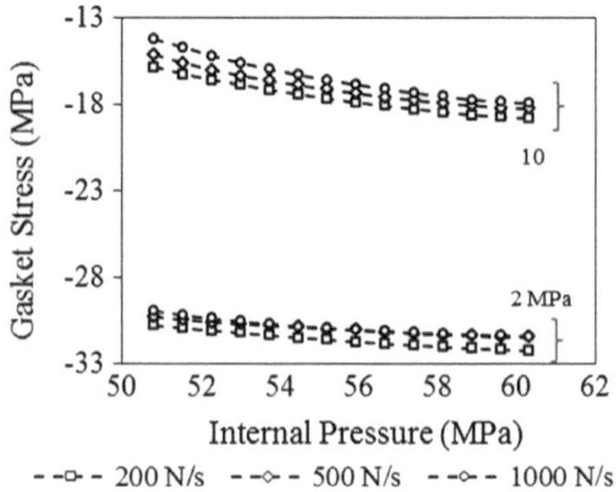

Fig. 6 Radial distribution of gasket stress for 120 kN total bolt preload

5.2 Sealing Performance

The sealing performance of flange joint is determined based on the gasket contact stress distribution. Leakage in gasketed joint depends on the gasket stress along inner circumference of gasket. The residual stress line given by Eq. (1) defines the minimum required compressive stress to be present on the gasket material as a residue, even after application of fluid pressure. While designing, the residual stress required to be on gasket depends on the internal pressure, which the flange joint should withstand. The gasket factor, m [14] for spiral wound gasket with graphite filler is 3. The residual stress line is a function of gasket material and fluid pressure.

The variation of gasket stress with internal pressure for different loading rates is shown in Fig. 7 for 120 kN bolt load. The intersection point for gasket stress variation line with residual stress line Eq. (1), is the minimum compressive gasket stress required on gasket and the corresponding internal fluid pressure is the safe value without leakage. The compressive stress on gasket increases with decrease in loading rate, i.e., the gasket gets more compressed, when loaded slowly. For 6 MPa internal pressure, the gasket stress at 200 N/s is −22.54 MPa, which reduces to −21.38 MPa at 1000 N/s. Thus, the safe internal pressure without leakage also reduces with increase in loading rate. The gasket stress during bolt up stage is almost same for different loading rate. The variation in the compressive stress increases with internal pressure. The safe internal fluid pressure for 200 N/s loading rate at 120 kN total bolt preload is 6.95 MPa and the same for 1000 N/s is 6.75 MPa (see Fig. 7).

$$Residual\,Stress\,Line\,(RSL) = mP \tag{1}$$

The analysis is carried out for different total bolt preloads viz., 120, 140 and 160 kN. The minimum leakage pressure for 80 mm NPS Class 600 flange joint under different bolt preload and varying loading rate is given in Table 2, along with percentage reduction in leakage pressure between 200 and 1000 N/s loading rate.

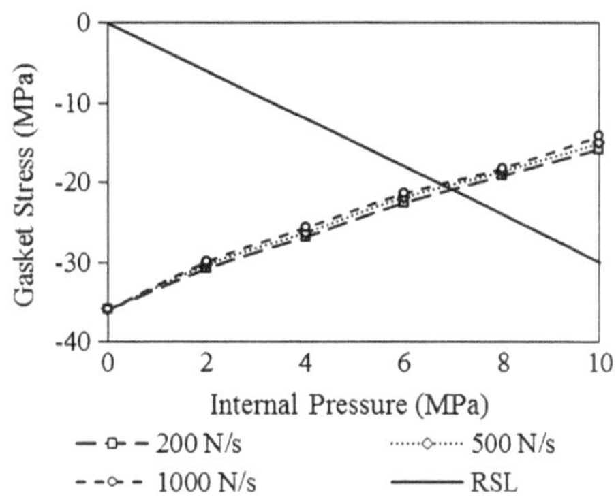

Fig. 7 Gasket stress distribution for different loading rate at 120 kN total bolt preload

Table 2 Leakage pressure for different loading rates and different bolt preload

Total bolt preload (kN)	Leakage pressure (MPa) at loading rates			% reduction between 200 and 1000 N/s
	200 N/s	500 N/s	1000 N/s	
120	6.95	6.82	6.75	2.96
140	8.20	8.00	7.95	3.14
160	9.30	9.11	9.00	3.33

At a particular bolt preload, the minimum leakage pressure decreases with increase in loading rate. For instance at 160 kN total bolt preload, the safe internal pressure is 9.30 MPa for 200 N/s loading rate, which reduces to 9.11 and 9.00 MPa for 500 and 1000 N/s loading rates respectively. The percentage reduction in safe internal pressure, between lower and higher loading rate increases with bolt preload, making the influence of loading rate more critical at higher bolt preload.

After gasket performance, more emphasis is given on the flange behaviour while designing flange joint. The flange behaviour is determined in terms of stress in flange ring, hub and fillet region between them. The stress in flange member depends on the stiffness of the joint, which in turn depends on gasket stiffness. The variation in flange stress is not substantial for different loading rate (deformation) of gasket, making it negligible.

6 Conclusions

Spiral wound gasket is tested under compression at different loading rates, to determine its deformation behaviour during loading and unloading. Substantial variation in deformation is observed, with comparatively higher deformation at lower loading rate. FE analysis is performed on flange joint including the gasket nonlinear characteristics at different loading rate. Based on the gasket contact stress variation, the leakage pressure is predicted in flange joint. The important observations are as following:

- The variation in gasket deformation behavior under compressive load is within 4% for the loading rates considered.
- The gasket compressive stress in flange joint is comparatively more at low loading rate. This results in withstanding higher internal fluid pressure for lower loading rate.
- The percentage reduction in safe internal pressure, due to different loading rate increases with bolt preload.
- The variation in flange stress is not significant for different loading rates of gasket.

The influence of gasket loading rate in sealing performance of the flange joint is highlighted. It has significant effect on the leakage pressure, due to variation in its compression behaviour with different loading rates.

References

1. Mathan G, Siva Prasad N (2010) Evaluation of effective material properties of spiral wound gasket through homogenization. Int J Press Vessel Pip 87:704–713
2. Nelson NR, Siva Prasad N, Sekhar AS (2016) Micromechanical modelling of spiral wound gasket under uniaxial compression. In: International congress on computational mechanics and simulation, pp 1116–1119
3. Fukuoka T, Takaki T (2003) Finite element simulation of bolt-up process of pipe flange connections with spiral wound gasket. J Pressure Vessel Technol 125:371–378
4. Nagy A (1996) Determination of the gasket load drop at large size welding neck flange joints in the case of nonlinear gasket model. Int J Press Vessels Pip 67:243–248
5. Nassar SA, Wu Z, Yang X (2010) Achieving uniform clamp load in gasketed bolted joints using a nonlinear finite element model. J Press Vessel Technol 132:031205
6. Drago J (2013) Initial gasket compression is key to safe, reliable flange joints. Sealing Technol 10–12
7. Estrada H (2015) Analysis of leakage in bolted-flanged joints using contact finite element analysis. J Mech Eng Autom 5:135–142
8. Bouzid AH, Galai H (2011) A new approach to model bolted flange joints with full face gaskets. J Press Vessel Technol 133:021203
9. Jenco JM, Hunt ES (2000) Generic issues effecting spiral-wound gasket performance. Int J Press Vessels Pip 77:825–830
10. Roos E, Kockelmann H, Hahn R (2002) Gasket characteristics for the design of bolted flange connections of MMC type. Int J Press Vessels Pip 79:45–52
11. Attoui H, Bouzid AH, Waterland JA (2016) On the buckling of spiral wound gaskets. J Press Vessel Technol 138: 041205-1-7
12. Nassar SA, Alkelani AA (2006) Effect of tightening speed on clamp load distribution in gasketed joints. SAE Technical Paper Series 2006-01-1250
13. ANSYS Mechanical APDL Element Reference (2012) ANSYS Inc., USA
14. Boiler and Pressure Vessel Code. Section VIII, Division I (2004) American Society of Mechanical Engineering

Effect of Bevel Angles on Tensile Strength of SS304 Steel Weld Joints

S.R. Pedapati, A.B. Sidiq, M. Awang, F.M. Hashim and S. Jebaraj

Abstract Welding is one of the most universal type of joining method used in the industry especially oil and gas industry. Good welding of pipe is crucial as bad welding can lead to cracks and failures. One of the parameter that affects the quality of weld is the bevel angle for single V butt joint weld. This paper aims to analyse the effect of different included bevel angle of SS304 stainless steel pipe weld. Simulation was performed in ANSYS which showed that 50-degree bevel angle is the optimum bevel angle for single V butt joint for SS304 steel with 13 mm thickness. To verify the simulation result, experiment was performed. The type of welding process that was used to perform the weld was Gas Tungsten Arc Welding (GTAW), also known as Tungsten Inert Gas (TIG) welding. The SS304 stainless steel was first sectioned and milled in order to achieve different bevel angle before being welded using TIG welding and machined again in order to perform tensile test. Tensile test was performed according to ASTM E8 standard in order to determine the strength of the welded pipe. Based on the values obtained from the tensile test such as Yield Strength and Ultimate Tensile Strength (UTS), the optimum bevel angle for single V butt joint is 50-degree.

1 Introduction

Welding is the joining of materials by means of heat and/or force with or without filler metal. In fact, welding is the only manufacturing process that has been used and applied in air, underwater and space [1]. One of the popular types of welding

S.R. Pedapati (✉) · A.B. Sidiq · M. Awang · F.M. Hashim
Department of Mechanical Engineering, Universiti Teknologi PETRONAS,
Bandar Seri Iskandar, Perak 32610 Malaysia
e-mail: srinivasa.pedapati@utp.edu.my

S. Jebaraj
Department of Mechanical & Manufacturing Engineering Technology,
Jubail Industrial College, Al Jubail 35718, Saudi Arabia
e-mail: zariah_j@jic.edu.sa

M. Awang (ed.), *2nd International Conference on Mechanical, Manufacturing and Process Plant Engineering*, Lecture Notes in Mechanical Engineering,
DOI 10.1007/978-981-10-4232-4_8

process is Tungsten Inert Gas (TIG) welding. TIG welding is widely used as one of the top choice of welding in fabrication industries, steel construction, maintenance and repairs. TIG is a type of arc welding wherein the weld is produced by using the heat produced by the arc between non consumable tungsten electrode and weld pool. In oil and gas industry, metal pipes are widely used to transport products and these pipes are welded together by welding processes such as TIG on single V butt weld.

However, welding on pipe can become problematic if not done properly. There are many parameters that affect the strength of the weld. Some researchers [2, 3] have studied the effect of welding parameters such as welding speed, welding current, plate thickness and welding voltage on the weld penetration and width. However, the geometry of the weld is also affected by the geometry of the pipe. Single V butt weld which is widely used for welding of pipe in oil and gas industry is affected by the bevel angle of the single V butt joint. However, not many researchers have been done on groove parameters like bevel angle in single V butt weld. Mahendramani [4] did a research closely related to groove angle and reported that angle in single V butt weld has a major effect on shrinkage of a metal and thus affecting the strength of the metal.

According to Kim et al. [5], producing good weld depends on the knowhow and long term experience. Trial and error is often used in order to achieve good weld. Improper welding work will result in failure of the plate such as crack on the weldment area which can lead to fracture and can be catastrophe. Weld that fail to attain standard during inspection will have to be welded again which is very costly. Some of the welding defects that are common in welding are porosity, wed splatter, poor fusion, shallow penetration and cracking which are due to poor welding parameters [6]. Murugan and Gunaraj [3] also explained the effects of welding parameters to the quality of weld.

One of the parameters that affect the strength of weld is bevel angle in butt weld [4]. Hamin [7] studied the effect of different bevel angles on the residual stress of the weld. It has been reported that the included bevel angle of single V butt weld affects the integrity of the weld joint in terms of the residual stress, distortion and the temperature profile of the weld. The paper also concluded that the optimum included bevel angle was 40-degree for A36 steel with 20 mm of thickness and 70 mm of width since it produced the least amount of distortion and residual stress. The current research on the other hand will determine the optimum bevel angle for single V butt joint in terms of the tensile strength of the weld by performing weld on metal with different bevel angle and later on performing the tensile test on the weld metal based on ASTM standard.

Thus, this research aims to analyse the effect of bevel angle in butt welded joint through TIG welding process. The main objectives of this research are to perform structural simulation using ANSYS on SS304 plate to determine the optimum bevel angle while determine the tensile strength of single V butt different bevel angles through experiment and to determine the optimum bevel angle for single V butt welds by validating the simulation with experiment results. This research covers shielded TIG welding process on SS304 stainless steel plate. The type of joint for the welding will be single V butt joint since double V butt joint is used for thicker metals. The only welding parameter that will be manipulated is the bevel angle of the metal to be welded (parent metal). The rest of the parameters will be kept constant in order to achieve good result. Tensile test according to ASTM E8 will be performed in order to analyze the strength of the weld. Tensile test specimen will be taken from welded SS304 plate using EDM wire cutting process.

Single V butt joint is used for relatively thinner materials since full penetration is not possible with thicker materials. For thicker materials, double V butt joined is used. The single V butt weld specimens are prepared as per American Welding Society (AWS) standards [8]. This was done so that the welding process will follow a recognized standard. AWS specifies the geometry of the welding plate and thus can be used for this project. As stated earlier, the type of welding process chosen is TIG welding process. Basically, it should be the same for TIG and MIG welding. AWS however does not discuss the effect of angle for single V joint. According to Murugan and Gunaraj [3] welding parameters can greatly affect the weld bead geometry such as weld penetration and width. This geometry can affect the strength of the weld. Murugan and Gunaraj [3] concluded that with increasing voltage, the width of the weld increases but the penetration of the weld decreases. However, they did not study the effect of angle of single V butt joint.

The thickness of the base metal is 13 mm and the root gap is from 0 to 3 mm. The root gap selected for this research is 2 mm. Figure 1 shows the nomenclature of bevel angle in V-butt joint. In order to determine the strength of the weld, tensile test according to ASTM A370 is conducted on the weld joint made of SS304 stainless steel [9]. This is due to the fact that ASTM standard is a widely recognized standard and is widely used in the oil and gas industry. The sample for tensile test must follow the dimension specified by ASTM. Basically, there are 3 types of specimen that can be selected for tensile test based on ASTM A370 which are standard specimen plate type, standard specimen sheet type and sub size specimen. For this research, sub size specimen is selected. The ultimate tensile strength for the material SS304 is 505 MPa while the Yield Strength is 215 MPa [10].

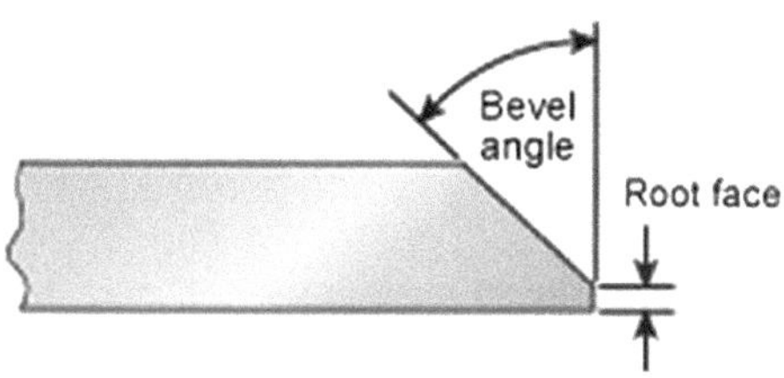

Fig. 1 Bevel angle nomenclature [11]

2 Experimental Procedure

The root gap of the butt joint kept constant at 2 mm. The thickness of the sheet is ½ inch which is the common thickness of plates and pipes. The dimension of the weld specimen is 250 × 60 × 13 mm. The bevel angles selected are 30°, 45°, 60°, and 75°. The angles are selected based on previous research by Hamin [7]. Most common bevel angle used for single V butt joint is 60°. Thus 60° bevel angle is also selected in the matrix.

2.1 Modelling

The first step is to select a reliable element type to perform structural analysis. Element type Solid 185 has been selected since it is used for 3D modelling of solid structures. Solid 185 is defined by 8 nodes, having 3 degrees of freedom at each node, that is in x, y and z directions. There are several material properties that need to be defined for the models. Each model is made up of SS304 steel as the parent metal and SS308L for the weld metal. There are 3 important parameters hat needs to be defined for each model, modulus of elasticity, Poisson ratio and density for both parent metal and weld area. Table 1 shows the properties of the two materials.

For the geometry of the model, there were 3 volumes created, first and second volumes are the parent metal and both are identical and should have the properties of SS304. Volume 2 is the weld area and consist of filler metal thus will have the properties of SS308L which is in fact the filler used in the real experiment. Each model has the same thickness, 13 mm, and the total width of the plate is 50 mm while the length is 100 mm. The modelling started with creating key points, followed by area from key points and lastly extruding the area in normal directions to create volumes. The final model for 50-degree bevel angle can be seen in Fig. 2.

Meshing was performed in ANSYS to represent field variables such as displacement by polynomial function that produces a displacement field compatible with applied boundary conditions. The size of the mesh has significant effect of the final result of the simulation, thus the size of the mesh was kept constant for all models. During meshing, the type of material properties was defined for each volume. Volume 1 and 2 were assigned to material 1 which has the properties of SS304 steel, while volume 3 was assigned material 2 which has the properties of SS308L. A fully meshed model can be seen in Fig. 3. The only load in this simulation is force. Force is applied on one end of the plate. The amount of force

Table 1 Material properties [10]

Material	SS304	SS308L
Modulus of elasticity, GPa	193	210
Poisson ratio	0.29	0.3
Density, kg/m^3	8000	8000

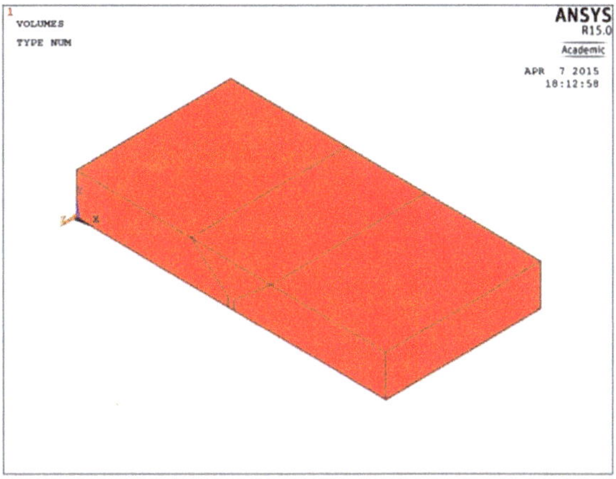

Fig. 2 50° bevel angle model 3.2.5 meshing

Fig. 3 Fully meshed 50° bevel angle model

applied was 50% of the force at Ultimate Tensile strength of SS304. The value of force applied was 2930803.5715 kN.

2.2 Experiment Methodology

The first process was cutting or sectioning the metal plate. The SS304 plate was cut in half producing 2 identical plates with dimension of 125 × 60 mm. This process was performed using Horizontal Band Saw machine. Lubricants were supplied through the process in order to minimize the heat generation during cutting which can affect the properties of the metal plate. The speed of depth of the hand saw machine was also set at minimum since stainless steel plate is a hard material compared to carbon steel and aluminium thus the speed of depth needs to be slowed down to prevent excess heat generation and machine failure.

After the metal plate has been sectioned, the side of the plate needed to be milled to produce bevel angle. The single V butt joint bevel angle had been milled according to test matrix. Manual milling machine had been used to introduce bevel angle. A root length of 3 mm was selected for each plate. Since manual milling machine by default is only able to mill flat surface thus to introduce angle to the metal plate, a special part known as angle plate was used.

After the SS304 plates have been milled according to test matrix, 2 identical plates in terms of bevel angle are welded using Tungsten Inert Gas (TIG) welding process. The current selected for the TIG welding process is 150 A [12]. In terms of the number of passes, for small bevel angle such as 30°, a single pass is sufficient to cover the single V butt joint. However, for large bevel angle, multiple passes were required in order to sufficiently weld the two metal plates together. In terms of the number of passes, for small bevel angle such as 30°, a single pass is sufficient to cover the single V butt joint. However, for large bevel angle, multiple passes were required in order to sufficiently weld the two metal plates together.

Deformation of the plate was noticed after the plates were welded together. This is due to the heat generated during welding and the metal plates were hit with hammer to straighten it. One of the most important aspects during welding is to ensure that the weld fully penetrates the weld plate. Only plate with 30° included angle did not achieve full penetration of weld. This is because the angle was too small, thus the space was not enough for the tungsten electrode to fully melt the parent metal and filler rod. The penetration can still be increased by using a wider root gap but this is not plausible for plate with 30° since 2 mm root gap has already been used which is very big. Figure 4 shows the back face showing full penetration. The colour around the welded region shows the heat affected zone.

The two plates welded together needed to be machined again (cut) in order to perform tensile test. The dimension of the samples needed to be machined was according to ASTM A307 [9]. The dimension according to ASTM A307 can be found in the literature review. The machine used to cut the tensile specimen from welded plates was Electrical Discharge Machining (EDM) wire cutting machine. In EDM wire cutting, electrical discharge was used to create sparks to cut the sample. The spark was created underwater, between the work piece and the wire that cuts the work piece. The EDM wire cutting machine produces very high quality products

Fig. 4 Back surface of welded plate showing full penetration

since it is precise because it uses AutoCad 2004. A model of the sample that needs to be cut from the work piece needs to be modeled in AutoCad 2004 which is connected to the machine. The model follows the dimension as per ASTM A370.

This specimen however was still not complying with ASTM A370 due to the thickness. The thickness of the specimen cut using EDM wire cutting is 13 mm whereas according to ASTM A370, the maximum thickness of the specimen for tensile testing for sub size specimen can only be 6 mm. Therefore, milling machine was used to reduce the thickness of the specimen. Simple horizontal milling was done on the specimen to reduce the thickness of the specimen from 13 to 6 mm. The dimension is shown in Fig. 5.

Tensile test was performed on each of the sample that had been welded and machined according to ASTM A307 [9]. The tensile test followed ASTM A307 standard. Tensile test machine was used to apply load on the sample until it fractures. The load caused the sample to be under tensile stress and elongate. The values of stress versus strain were plotted in order to determine the mechanical properties of the sample that has been welded such as the yield stress and ultimate tensile strength. Before tensile test can be performed, 5 thickness reading and 5 width reading of the gauge area needs to be measured and the average needs to be calculated for each sample as shown in Table 2. The thickness of the gauge area of each sample was measured using micro-meter screw gauge while the width was measured using Vernier calliper. Both equipment gave automatic reading.

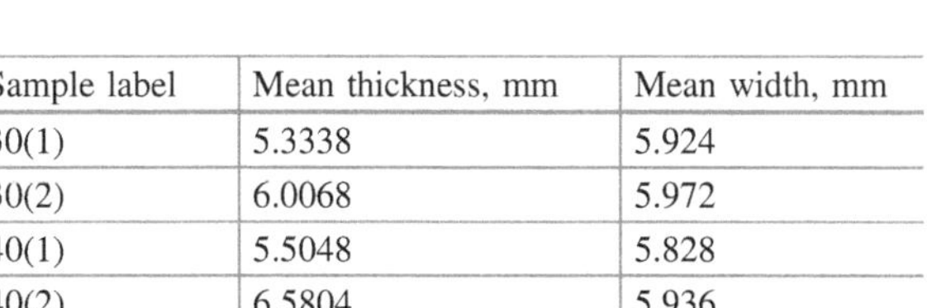

Fig. 5 Dimension of final tensile test specimen

Table 2 Thickness and width of tensile test sample

Sample label	Mean thickness, mm	Mean width, mm
30(1)	5.3338	5.924
30(2)	6.0068	5.972
40(1)	5.5048	5.828
40(2)	6.5804	5.936
50(1)	5.8020	5.976
50(2)	6.0660	5.832
60(1)	5.5370	5.844
60(2)	5.6970	5.730
70(1)	4.9140	5.810
70(2)	5.1690	5.684

3 Results and Discussion

3.1 Results

The final step of the project was to analyze the tensile test result obtained from sample with different bevel angle. The properties obtained from tensile test were analyzed and optimum bevel angle for single V butt joint weld for ASTM A106 Grade B pipe were determined. The simulation performed was structural simulation whereby force was applied on each model having different bevel angle. The value of force applied on every model was the same in order to determine the model sustaining the least amount of stress. Table 3 shows the maximum stress sustained in X direction for each model. Summary of Ultimate Tensile Strength and yield stress for all the samples and average yield stress for each bevel angle is shown in Table 4. The fractured tensile test specimen is shown in Fig. 6.

Not all plates that were welded showed good weldment. 30-degree bevel angle plate did not achieve full penetration. The root area was not welded and remained empty. The filler metal did not reach the root area and the electrode was not able to reach deep enough to melt the parent metal and filler metal to fill in the root area.

Table 3 Maximum stress sustained in X axis

Model type	30°	40°	50°	60°	70°
Maximum stress sustained, X axis, MPa	892.90	773.77	659.62	927.09	1537.8

Table 4 Summary of ultimate tensile strength, yield stress and average yield stress

Sample	S1 UTS, MPa	S2 UTS, MPa	Avg UTS, MPa	S1 yield stress, MPa	S2 yield stress, MPa	Avg yield stress, MPa
30°	514.976	499.152	507.064	320.599	274.634	297.616
40°	594.95	598.429	596.689	342.821	330.544	336.682
50°	633.512	637.191	635.351	380.683	390.699	385.691
60°	615.557	614.72	615.138	367.580	366.061	366.820
70°	628.908	609.157	619.032	372.305	347.865	360.085

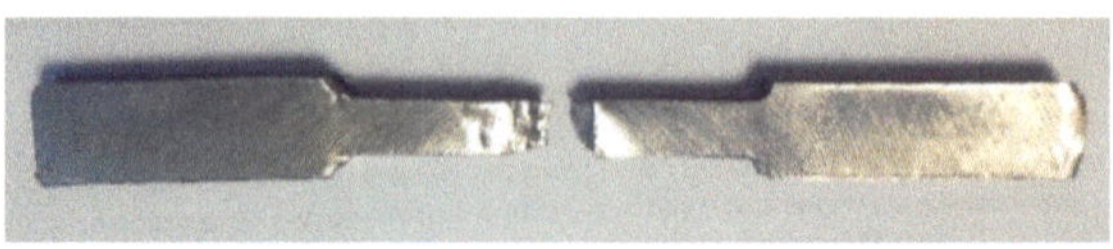

Fig. 6 Fractured tensile test specimen

3.2 Discussion

As a preliminary study, simulation was performed in ANSYS. The simulation involved applying force 2.9308035715 MN which is 50% of the force that the sample sustains at the ultimate tensile stress (505 MPa) for SS304 material. The trend for the maximum stress sustained is similar even if the value of stress is increased up to 100%. It should be noted that this stress value is the maximum stress sustained when the force for all the samples are similar. Thus, a sample with lower stress is better compared with a sample with higher stress. This is because, for the same amount of force, if a model experiences higher stress, then the model will fail first compared to the model which experiences lower stress for the same amount of force applied provided that both models are made up of the same material which is true for this simulation.

As can be seen in Fig. 7, the model with 50° bevel angle sustains the least amount of stress for the same value of force, thus 50° bevel angle is the optimum angle. Thus the value is matching with experimental data. Stress in X direction is considered instead of Von Mises due to the fact that the force applied is only in one direction, the x axis. Thus, stress analysis in x direction is sufficient. Von Mises stress is important when the model is subjected to force in multiple directions. There are 2 types of stress versus strain graph, one using engineering stress and another using real stress. The difference is in area, in real stress the instantaneous area is taken while in engineering stress, original area is considered to calculate stress. This is the reason why the stress value in engineering stress versus strain graph decreases after UTS since the area actually reduces and the sample actually should have more stress since its area is in fact reduced. However, engineering stress is used in engineering application since after necking; the material is no longer safe and usable. The result of the experimental tensile test is based on engineering stress versus strain.

Based on Fig. 8, it can be seen that the UTS value increases as the bevel angle is increased until the bevel angle reaches 50°. 50-degree sample achieves the maximum UTS. The UTS value reduces as bevel angle is increased further and remains almost the same for 60° and 70°. This shows that the SS304 plate with 50° bevel angle is the optimum angle since it is able to hold the maximum amount of load before it fractures. In tensile test, the force applied is in tension thus the UTS value reflects the maximum tension the sample is able to hold before it starts to fracture and break.

Fig. 7 Maximum stress sustained in X axis versus bevel angle

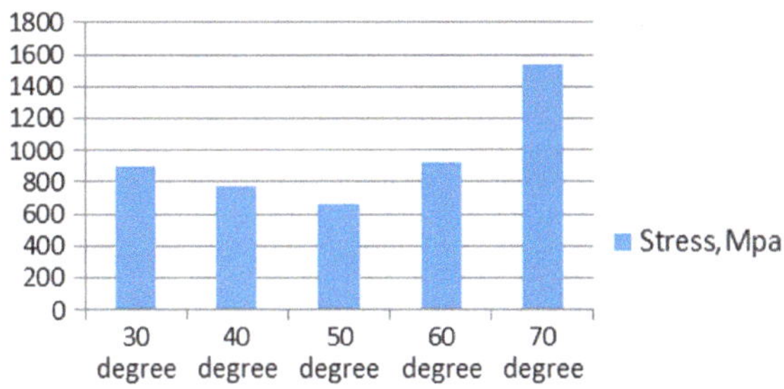

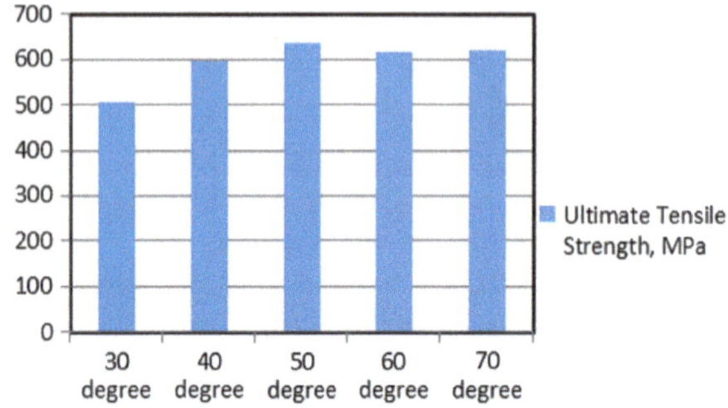

Fig. 8 UTS versus bevel angle

The metal plate welded with 30° bevel angle shows the weakest UTS value which is expected even after the metal was welded. This is because there was not enough melting and fusion between the parent metal and the weld filler. The reason is because bevel angle of 30° does not have enough space for the electrode to go close to the grove area thus only the top part of the parent metal receives enough heat and melts while the bottom part of the parent metal does not melt nor does the filler metal reaches the bottom part of the parent metal (root length section). The plate welded with 30° bevel angle also shows that there is no full penetration during welding which proves that bevel angle of 30° does not have enough space for perfect welding.

The ultimate tensile strength value increases as the bevel angle is increased from 30°, 40° and 50°. This is due to the fact that as the bevel angle is increased, the tungsten electrode is able to reach deeper into the bevel angle and thus it is able to melt even the bottom part of the parent metal. This also causes the metal plate to experience full penetration which increases the strength of the metal plate.

The ultimate tensile strength for bevel angle 60° and 70° is almost similar which are higher than 40° and 30°. However, the bevel angle 60° and 70° are not practical to be used in the industry since the bevel angle is too larger. Since the bevel angle is too large, it will require a lot of filler metal to fill the gap which can increase the cost. The large bevel angle will also require more number of passes in order to fill up the gap. For 70° bevel angle, 3 passes were insufficient to completely fill up the gap. If the gap is not completely filled up, then the weld area can become stress concentrator due to its reduced area. The cross sectional area of the weld will be lower compared to the rest of the structure or pipe, and thus it will experience larger stress compared to the rest of the structure or pipe and will most likely fail even at load that the rest of the pipe is able to sustain.

3.2.1 Yield Stress

Apart from ultimate tensile strength, another important result that can be obtained from stress versus strain graph taken from tensile test is the 0.2% offset Yield Stress. Yield stress or also known as Yield strength is the maximum stress the

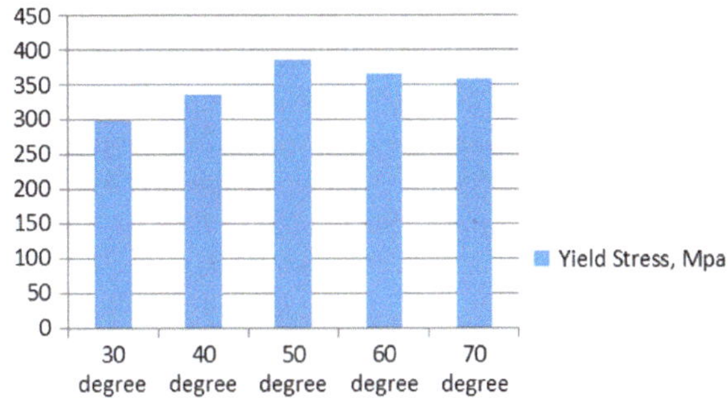

Fig. 9 Yield stress versus bevel angle plot

sample is able to hold before the material starts plastic deformation. Stress lower than the yield stress will result in elastic deformation were by the material will be able to revert back to its original dimension and microstructure after the force is removed. Therefore, the material will only sustain temporary deformation. Therefore, the material will only sustain temporary deformation.

However, if the material experiences stress above the yield stress, the material will experience permanent deformation. The material will not be able to revert back to its original dimension and microstructure even after the load is removed. Strain hardening will occur in permanent deformation which will increase the tensile strength but it will cause the material to become less ductile. In most engineering application, design of structures is made based on the yield strength. The structure is assumed to fail if the stress exceeds the yield stress.

The trend for yield stress follows the same trend for ultimate tensile strength for every degree of bevel angle as shown in Fig. 9. The SS304 plate with 50° bevel angle is still the optimum bevel angle based on the highest 0.2% offset yield stress which is 385.691 MPa. However, it should be noted that for yield stress bevel angle 60° is better than 70° since it has slightly higher yield stress although 70° bevel angle has higher ultimate tensile strength compared to 60° bevel angle.

4 Conclusions

The simulation has been successfully performed in ANSYS. Five models have been created, each for every test matrix. Structural load (force) has been applied on every model in order to determine the stress sustained by the models. The simulation showed that for the same value of force applied, the model with 50° bevel angle single V butt joint experienced the smallest value of stress.

In order to validate the simulation, experiment has been performed. Stainless steel SS304 plates with 13 mm thickness has been successfully machined and welded with different bevel angles according to test matrix. Two samples have been successfully cut from each welded plate and tensile test has been performed to determine the tensile strength (Ultimate Tensile Strength and Yield Stress).

In order to validate the experiment, simulation has been performed in ANSYS software according to the test matrix for each bevel angle. A total of 5 models have been tested, each model having different bevel angles.

Based on the experimental result, the optimum bevel angle for butt joint of 13 mm thickness SS304 steel is 50° since it has highest UTS value compared to the rest of the bevel angles. Steel plate with bevel angle 50° also resulted in the highest yield stress compared to the rest of the samples, thus it is the most suitable bevel angle for butt joint of SS304 steel and the simulation result has been validated.

Acknowledgements The authors acknowledge the support provided by the Universiti Teknologi PETRONAS, Malaysia for this research under URIF grant (0153AA-B71).

References

1. Patel PD, Patel SP (2011) Prediction of weld strength metal active gas (MAG) welding using artificial neural network. Int J Eng Res Appl (IJERA) 036-044
2. Srinivasa Rao P, Gupta OP, Murty SSN (2005) Influence of process parameters on bead geometry in pulsed gas metal arc welding. IIW International Congress, Mumbai India
3. Murugan N, Gunaraj V (2005) Prediction and control of weld bead geometry and shape relationships in submerged arc welding of pipes. J Mater Process Technol 168(3):478–487
4. Mahendramani G (2012) Effect of included angle in V-groove butt joints on shrinkages in submerged arc welding process. Int J Eng Sci Technol (IJEST) 14
5. Kim IS, Jeong YJ, Lee CW, Yarlagadda PKDV (2003) Prediction of welding parameters for pipeline welding using an intelligent system. 22:713–719. doi:10.1007/s00170-003-1589-y
6. Lincoln Electric (1994) The procedure handbook of arc welding. Lincoln Electric, Cleveland, Ohio. ISBN 99949-25-82-2
7. Hamin MZ (2013) Finite element analysis of residual stress in welded thick plates of a single butt weld with different included bevel angles. UTPEDIA. Retrieved 10 Dec 2014 from http://utpedia.utp.edu.my/13733/1/completed.pdf
8. American Welding Society (AWS), standard welding preparation for single V butt weld, AWS D1.1: structural welding code-steel. Retrieved 8 Dec 2014 from https://law.resource.org/pub/us/cfr/ibr/003/aws.d1.1.2000.pdf
9. ASTM A370-14, Standard Test Methods and Definitions for Mechanical Testing of Steel Products, ASTM International, West Conshohocken, PA, 2014, www.astm.org
10. AISI Type 304 Stainless Steel, ASM Aerospace Specifications Metal Inc. http://asm.matweb.com/search/SpecificMaterial.asp?bassnum=MQ304A
11. Technical knowledge, design part 3. The Welding Institute. http://www.twi-global.com/technical-knowledge/job-knowledge/design-part-3-092
12. Smith LD (2001) The fundamentals of gas tungsten arc welding: preparation, consumables, and equipment necessary for the process. The fabricator. Retrieved 8 Dec 2014 from http://www.thefabricator.com/article/arcwelding/the-fundamentals-of-gas-tungsten-arc-welding–preparation-consumables-and-equipment-necessary-for-the-process

Developing a Finite Element Model for Thermal Analysis of Friction Stir Welding by Calculating Temperature Dependent Friction Coefficient

B. Meyghani, M. Awang, S. Emamian and Nor M. Khalid

Abstract One of the main sources of the heat generation during the Friction Stir Welding (FSW) process is the friction force. There is a need to define the friction coefficient in order to accurately predict the simulated model. Many authors assumed the friction coefficient as a constant value. However, such assumptions may affect the reliability of the results obtained. In this paper, ABAQUS® software is used to simulate thermal behavior during the FSW process. In order to calculate temperature dependent friction coefficient values, the Coulomb friction law is modified in MATLAB®. The results show that the heat generation strongly depends on the motion of the welding tool which is a combination of rotational and transverse speeds. Furthermore, the rise of the rotational speed increases the welding temperature, because of the higher friction heating, higher stirring, and higher mixing of the material. In contrast, the increase of the transverse speed decrease the welding temperature. The results also demonstrate that the temperature profile during the welding is asymmetrically distributed in the welding cross section (temperature being higher on the advancing side). Experimental measurements have been conducted to validate the temperature profile. In conclusion, the simulation results are consistent with the experimental data as well as published results.

Keywords Friction Stir Welding (FSW) · Friction coefficient · Finite Element Model (FEM) · Coulomb friction law

B. Meyghani · M. Awang (✉) · S. Emamian
Department of Mechanical Engineering, Universiti Teknologi PETRONAS, 32610 Seri Iskandar, Perak, Malaysia
e-mail: mokhtar_awang@utp.edu.my

N.M. Khalid
Friction and Forge Processes Group, Joining Technologies Group, TWI Ltd, Cambridge, UK

M. Awang (ed.), *2nd International Conference on Mechanical, Manufacturing and Process Plant Engineering*, Lecture Notes in Mechanical Engineering,
DOI 10.1007/978-981-10-4232-4_9

1 Introduction

Investigation of the thermal behaviour during the Friction Stir Welding (FSW) is crucial for increasing the quality of the weld. Basically, FSW involves a complex physics phenomena caused by the severe plastic deformation and the complicated friction behaviour [1]. Due to the complexity of the process, studying in depth by experimental methods is challenging, hence, to investigate the leading mechanisms and solve the process governing equations; Finite Element (FE) modelling attract enormous research interest. It should be noted that, one the main sources of the heat generation during the FSW is the friction force. Therefore, a proper definition of the friction coefficient in the FE model is significant.

Basically, FE modelling of the FSW process separates into two different levels. Investigation of the effect of the welding parameters, contact mechanisms and geometry near the pin-tool area (Heat Affected Zone) is called the local level, while the analyses of the entire workpiece structural such as temperature distortion and residual stresses is called global level. In order to investigate the process at the different levels, three different methods are proposed, Lagrangian, Eulerian and Arbitrary Lagrangian–Eulerian (ALE).

In the Lagrangian approach the domain is separated into elements and nodes, which will distort and travel with the deforming material (suitable for global level modelling). Whereas, in the Eulerian formulation, the nodes and elements remain constant and the material is able to flow through the domain. The ALE method is considered as a suitable method for local level modelling, because the computational mesh inside the domain is able to move arbitrarily to optimize the shapes of the elements. In this method, the mesh at the boundaries and interfaces of the domains can move along with the materials to precisely track the boundaries and interfaces of a multi-material system. A finite element volume code was used by Askari et al. [2] as an advection-diffusion solution in order to solve the energy balance equation. In the model the elastic response was taken into account. Thus, the strain rate, strain, the material flow around the tool and temperature distribution in the model could be predicted successfully. However, in the research a simplified friction model was considered in which the friction coefficient was assumed as a constant value. Mixing fraction was also used in the model to determine the ratio of the material flow from the advancing side to the retreating side. Results showed that the material movement started from the advancing side, and the flow of the material in the advancing side is more than the retreating side. Moreover, the temperature was observed higher in the advancing side. In the meantime, a friction base model was proposed by Assidi [3] using both Eulerian and ALE formulation. In the model the friction coefficient was assumed to be a constant value. By using ABAQUS®, Xu and Deng [4] employed a three-dimensional FE model in order to simulate the material flow characteristics and the equivalent plastic strain distribution during the FSW process. The ALE formulation, adaptive meshing, large elasto plastic deformations and temperature dependent material properties were used to increase the model accuracy. The velocity gradient near the pin was also recorded. However, the model was not a fully coupled thermomechanical model, because the temperature was defined as an input parameter which was obtained from experimental measurements. In addition, a constant friction coefficient value was used. Schmidt and Hattel [5] employed a three dimensional fully coupled thermomechanical FEM. In the model, a constant friction coefficient value and Johnson–Cook hardening law were used. Based on the estimation of the error, an adaptive re-meshing structure and ALE formulation as a splitting method were used to predict the temperature distribution. Results showed that in the advancing side, the material velocity and the temperature are higher than in the retreating side. Santiago et al. [6] used the ALE method in a simplified friction model in which a partial sliding/sticking condition at the pin shoulder/work-piece interfaces were considered. By assuming a sub-grid stabilized mixed velocity, pressure and thermal behaviour of FSW were investigated in coupled thermo-rigid-plastic models [2, 7–10]. Eulerian and Arbitrary Lagrangian Eulerian (ALE) methods were used in all of the models. Besides that, by using constant friction coefficient values, Dialami et al. [10–12] and Chiumenti et al. [7, 13] established a suitable kinematic framework for modelling the FSW processes and comparing the solid method with others methods. In these models, a mixture of ALE, Eulerian and Lagrangian methods at different welding zones with an efficient coupling strategy was presented. Results showed that, the kinematic setting permitted efficient treatment of arbitrary pin geometries and facilitated the application of boundary conditions.

Although, many studies are conducted to improve the accuracy of the process modelling, however, there are still some unsolved challenges in the terms of the selection of an effective contact model, and also the prediction of the temperature dependent friction coefficient values by using straightforward methods. As discussed in the work done by researchers [7, 14, 15] the friction coefficient seldom considered as a significant factor. They usually used a limited contact formulation in which the friction coefficient was assumed to be constant. However, in reality, the friction coefficient is a function of velocity, temperature, and strain rate, therefore the constant values of the friction coefficient during different welding conditions and different welding steps resulting the termination of the simulation. To explain the problem, the shear stress, which is created by the frictional force cannot be reached more than the upper limit of the material shear failure. In addition, in sliding condition, using a constant value of the friction coefficient in Coulomb friction law shows a poor representation of the friction phenomena. In this regard, the modified version of the Coulomb friction law in which the coefficient depends on the temperature can increase the accuracy of the simulation. Therefore, an accurate implementation of the contract condition in the model such as calculation of the suitable values for the friction coefficient is an appropriate method for solving complicated governing equations.

Therefore, in this paper in order to obtain more accurate and simple prediction of the temperature distribution, thermal history and temperature profile, a mathematical formulation is used to calculate the temperature dependent values of the friction coefficient and the slip rate in a partial sliding and sticking condition. Then, the calculated values are employed in order to develop a three-dimensional FE model of FSW of AA6061-T6 at various welding speeds. Finally, the capability of the mathematical formulation and the FE model is verified by comparing numerical results with experimental measurements and published results.

2 Mathematical Model Description

2.1 *Sliding, Sticking and Partial Sliding/Sticking Condition*

According to the Coulomb friction law, the shear stress at the interfaces is expressed as,

$$\tau_{fric} = \mu P_0 \quad (1)$$

where τ_{fric} is the friction shear stress, μ is the friction coefficient and P_0 is the axial contact pressure.

With the increase of the temperature, von Mises shear stress criterion dominates the plastic deformation. The shear stress (τ_y) in von Mises theory is calculated as follows,

$$\tau_y = \frac{\sigma_y}{\sqrt{3}} \quad (2)$$

where σ_y is the yield stress.

In the FEM, the strain rate and temperature are two important parameters for modelling the plasticity behaviour. In order to avoid large plastic deformation problems and the distortion of mesh the elastic-plastic Johnson-Cook material law is used as follows,

$$\sigma_y = [A + B(\varepsilon_P)^n]\left[1 + C\left[\frac{\dot{\varepsilon}_P}{\dot{\varepsilon}_0}\right]\right]\left[1 - \left[\frac{T_{FSW} - T_{room}}{T_{melt} - T_{room}}\right]^m\right] \quad (3)$$

where ε_P, $\dot{\varepsilon}_P$ and $\dot{\varepsilon}_0$ are the effective plastic strain, the effective plastic strain rate and the normalizing strain rate respectively. The input parameters of Johnson-Cook model for Al 6061-T6 are explained in Table 1.

Table 1 Johnson-Cook material law parameters

Room temperature T_{room} (°C)	25
Melting temperature T_{melt} (°C)	582
Yield stress (A) dimensionless	546
Strain factor (B) dimensionless	678
Strain exponent (n)	0.71
Strain rate factor (C) dimensionless	0.024
Temperature exponent (m)	1.56

In full sliding ($\delta = 0$) and full sticking ($\delta = 1$) conditions, the shear stress in each condition can be written as follows,

$$\tau_{fric} = \mu P_0, \quad when \quad \delta = 0 \tag{4}$$

$$\tau_{fric} = \tau_y, \quad when \quad \delta = 1 \tag{5}$$

where μ is the friction coefficient, τ_y is the shear stress and P_0 is the axial pressure at the shoulder bottom and the pin bottom which is calculated as,

$$P_0 = \frac{F_N}{\pi R_S^2} \tag{6}$$

where F_N is the normal force and R_S is the shoulder radius.

The contact shear stress at the shoulder bottom and the pin bottom surfaces in a partial sliding/sticking condition can be written as,

$$\tau_0 = \delta\tau_y + (1 - \delta)\mu P_0 \tag{7}$$

where δ is the slip rate,

The shear stress for the pin side area can be defined as,

$$\tau_1 = \mu P_0 \sin\alpha \quad when \quad \delta = 0 \tag{8}$$

$$\tau_1 = \tau_y \quad when \quad \delta = 1 \tag{9}$$

The shear stress for the pin side area in a partial sliding/sticking condition can be explained as,

$$\tau_1 = \delta\tau_y + (1 - \delta)\mu P_0 \sin\alpha \tag{10}$$

As can be seen in Fig. 1, for a conical pin profile, α is the cone angle and for a threaded cylindrical pin profile α is the thread angle.

Due to the larger geometry in the shoulder bottom and the pin bottom areas, the shear stress amount at the pin side area is less than the above-mentioned areas. By employing Eqs. 7 and 10 and solving a linear system of two equations with two variables, the value of the friction coefficient can be calculated as,

$$\mu = \frac{\tau_0 - \tau_1}{(1 - \delta)P_0(1 - \sin\alpha)} \tag{11}$$

By incorporating Eq. 11 into Eq. 7 or Eq. 10 the value of the slip rate can be achieved as follows,

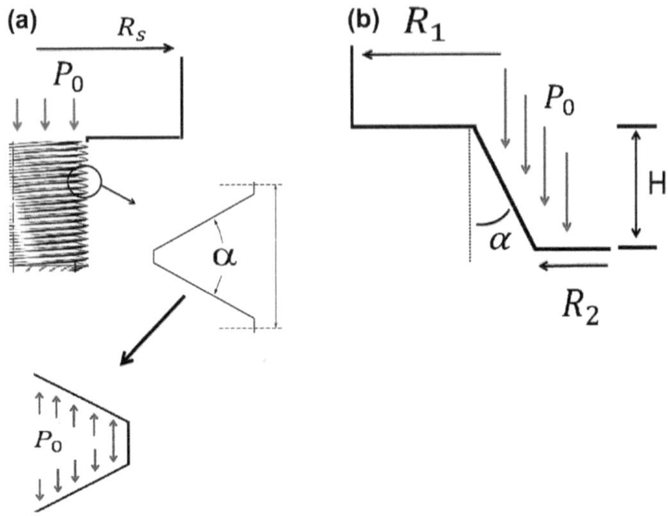

Fig. 1 Pressure profile at the tool interface **a** threaded cylindrical **b** conical

$$\delta = \frac{\tau_1 - \tau_0 \sin\alpha}{(1 - \sin\alpha)\tau_y} \quad (12)$$

Therefore, the final equation for calculating the value of the friction coefficient is defined as follows,

$$\mu = \frac{\tau_0 - \tau_1}{\left(1 - \frac{\tau_1 - \tau_0 \sin\alpha}{(1-\sin\alpha)\tau_y}\right) P_0 (1 - \sin\alpha)} \quad (13)$$

The governing equations need to be solved to derive the value of the friction coefficient and the slip rate. In the first step, the ratio of the shear stress (β) can be expressed as follows,

$$\beta = \frac{\tau_0}{\tau_1} \quad (14)$$

The shear stress rate can be obtained by resolving Eqs. 7, 10 and 14. Since Eqs. 7 and 10 are in the form of $ax + by = c$ and $a'x + b'y = c'$ (where, a, b, c, a', b' and c' are nonzero numbers), these two equations can be assumed as a system of two linear equations with two variables. The system assumptions are explained in Table 2.

Table 2 The system assumptions

a	b	c	a'	b'	c'	x	y
τ_y	1	τ_0	τ_y	$\sin\alpha$	τ_1	δ	$(1 - \delta)\mu P_0 \sin\alpha$

In the system $\frac{a'}{a} = 1$ and $\frac{b'}{b} = \frac{1}{\sin\alpha}$, which means, $\frac{a'}{a} \neq \frac{b'}{b}$, therefore the system has one solution,

$$1 < \beta < \frac{1}{\sin\alpha} \quad \alpha > 0 \tag{15}$$

Based on the assumptions made the value of the shear stress ratio and Alfa, which are considered as 1.5 and 30 °C respectively, therefore,

$$\beta = \frac{\tau_0}{\tau_1} = 1.5 \tag{16}$$

By putting Eqs. 6, 12, 13 and 16 into Eq. 7, the value of the shear stress at the shoulder bottom and the pin bottom can be written as follows,

$$\tau_0 = \left(\frac{\frac{\tau_0}{1.5} - \tau_0 \sin\alpha}{(1 - \sin\alpha)}\right) + \left(1 - \left(\frac{\frac{\tau_0}{1.5} - \tau_0 \sin\alpha}{(1 - \sin\alpha)\tau_y}\right)\right)\left(\frac{\tau_0 - \frac{\tau_0}{1.5}}{\left(1 - \frac{\frac{\tau_0}{1.5} - \tau_0 \sin\alpha}{(1 - \sin\alpha)\tau_y}\right)\left(\frac{F_N}{\pi R_S^2}\right)(1 - \sin\alpha)}\right)\left(\frac{F_N}{\pi R_S^2}\right) \tag{17}$$

where the input parameters are α, F_N, R_s and τ_y.

As mentioned earlier, in order to calculate the shear stress (τ_y), the value of the yield stress (σ_y) need to be calculated by resolving of the Johnson-Cook equation (Eq. 3).

3 Experimental Set-up

AA6061-T6 alloy is chosen for the work piece material while steel H13 is used for the tool. The dimension of the workpiece before the weld was 10 mm in thickness; 200 mm in length and 100 mm in width. The diameter of the shoulder was 18 mm and the length of the pin was 6 mm. The FSW-TS-F16 machine was employed to do the experiments. The plunge depth of the shoulder into the workpiece was set to 0.15 mm. Table 3 illustrates different welding speeds which were used for the experimental tests.

Moreover, thermal cycles during the welding were measured by using K type thermocouples. To measure and record the temperatures during the welding, at various locations holes with a depth of 3 mm on the top surface of the workpiece,

Table 3 Different welding parameters

Process parameters	Values
Axial force (N)	5000
Rotational speeds (RPM)	800, 1200, 1600
Transverse speeds (mm/min)	40, 70, 100

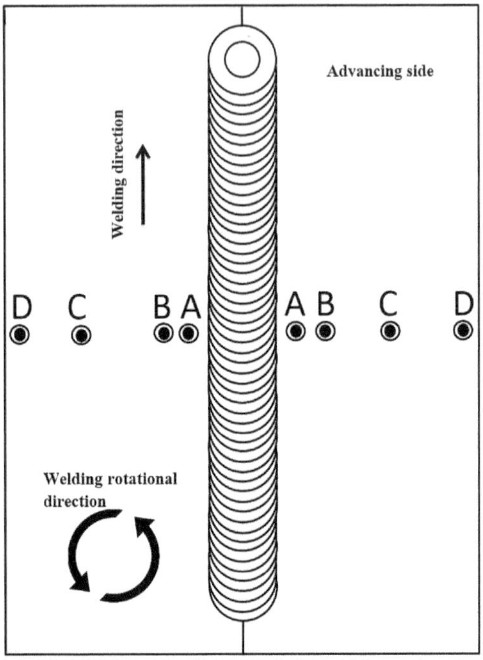

Fig. 2 Thermocouples positions at the work piece, 2 mm (point A), 4 mm (point B), 10 mm (point C) and 40 mm (point D) away from the welding path at both advancing and retreating sides

as shown in Fig. 2 were punched at different locations which were 2 mm (point A), 4 mm (point B), 10 mm (point C) and 40 mm (point D) away from the welding seam at both sides of the weld (advancing and retreating sides).

4 Numerical Model Description

In order to create a verifiable prediction of temperature and displacement, a three-dimensional FEM is established. The ABAQUS® commercial software is used for modelling of the process, as ABAQUS® utilises the Newton method as a backward difference scheme and uses a coupled system for solving both temperature and displacement equations at every time increment and each node; therefore, it is suitable for modelling of the FSW process. As can be seen in Fig. 3 there is a region that has a coarse mesh which can be used to minimize the simulation time. In addition, to reduce the CPU time without compromising the model accuracy, the Arbitrary Lagrangian Eulerian (ALE) approach, the adaptive meshing and the mass scaling techniques are applied. The work piece is partitioned into 12 areas which are constrained in the Y direction. A general purpose linear brick element called coupled temperature displacement (including both specified degrees of freedom) with reduced integration (C3D8RT) is used, because both temperature and displacement degrees of freedom have to be considered in relation to the severe plastic deformation.

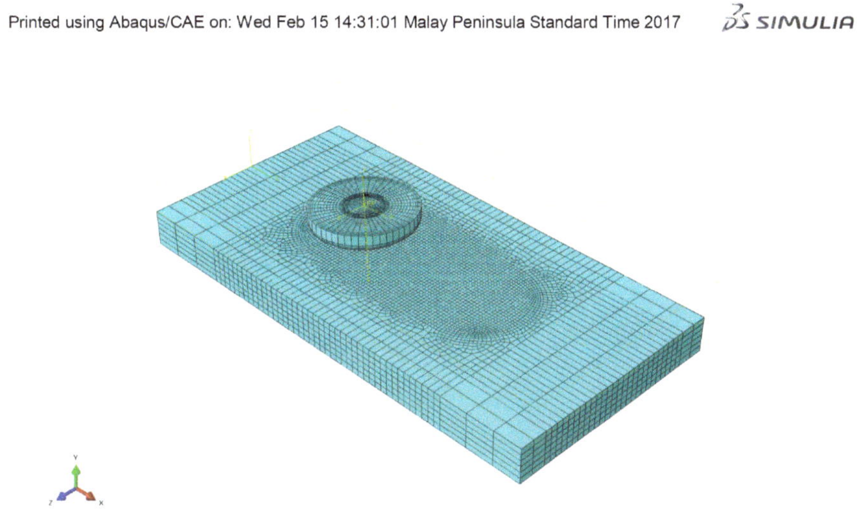

Fig. 3 FE model of FSW, mesh system, tool geometry and tool coordinate system

Table 4 Mechanical properties of AA6061-T6 (temperature dependent)

Temp (°C)	Young's modulus (GPa)	Thermal conductivity (W/mK)	Poisson's ratio
148.9	88.54	162	0.34
204.4	66.19	177	0.34
260	63.09	184	0.34
315.16	59.19	192	0.34
371.1	53.99	201	0.34
426.7	47.47	207	0.34
148.9	40.30	217	0.34
204.4	31.72	223	0.34

Mechanical properties and thermal conditions of the model are presented in Tables 4 and 5. The dimensions of the work piece in the numerical model are the same as experimental dimensions and the tool is modelled as a rigid body without thermal degrees of freedom.

5 Results and Discussion

5.1 The Effect of the Temperature in the Friction Coefficient

When the values of the friction coefficient at different temperatures are scarcely available in the model, an accurate calculation of the heat (generated by the friction

Table 5 Temperature dependent material properties

Temp (° C)	Coefficient of thermal expansion (°C)	Specific heat capacity (J/Kg °C)	Density (kg/m^3)
37.8	2.345×10^{-5}	95	2685
93.3	2.461×10^{-5}	978	2685
148.9	2.567×10^{-5}	1004	2667
204.4	2.669×10^{-5}	1028	2657
260	2.756×10^{-5}	1052	2657
315.6	2.853×10^{-5}	1078	2630
371.1	2.957×10^{-5}	1104	2630
426.7	3.071×10^{-5}	1133	2602

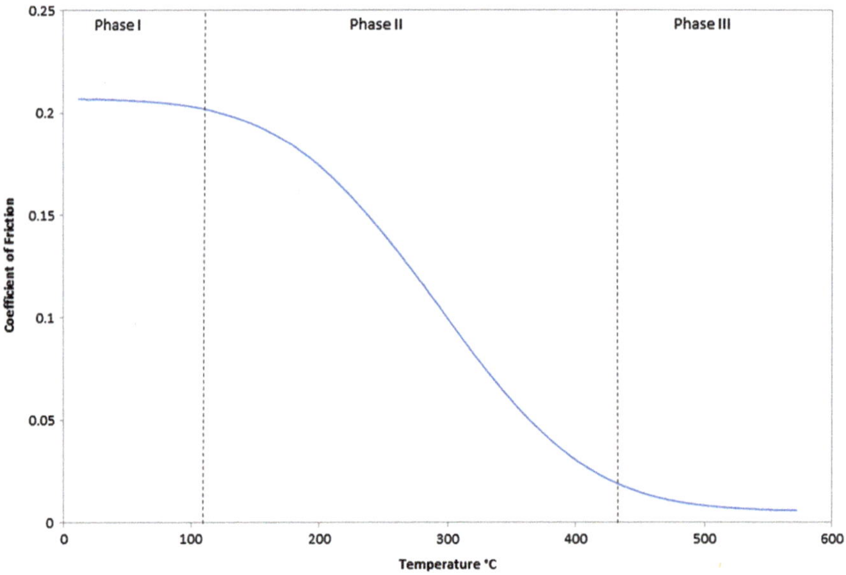

Fig. 4 Calculated values of the friction coefficient

force) is difficult. From the calculated results, which are shown in Fig. 4, the friction coefficient for AA6061-T6, varies within the range of 0.207089 to 0.000582 when the temperature is in a range of 25 °C (the room temperature) to 580 °C (the material melting temperature). As can be seen in Fig. 4 the behaviour of the friction coefficient is separated into three phases.

The values for the friction coefficient are approximately constant, when the welding temperature is less than 120 °C, because the material has not yet completely deformed. In the study that is done by Guide et al. [16], based on dry sliding tests in a full sliding condition, temperature dependent values of the friction coefficient were found approximately around 0.25 at room temperature for aluminium/steel. However, in sticking condition, where the friction force exceeds the shear yield strength of the workpiece, the coefficient is approximately estimated around 0.5. Further studies on the temperature dependent friction coefficient values were retrieved from Zhang et al. study [17]. Different relative surface velocities in different temperatures for 3003-H18 aluminium alloy were used to observe the friction behaviour. The study found that the friction coefficient was 0.274 at room temperature.

Therefore, during the FSW process the temperature range inside the nugget zone varies in a very large range (60–90% of the material melting point between 350 and 450 °C) [18]. This fluctuation in the temperature caused a critical change in the mechanical features of the material in the nugget zone, the contact state of the tool-workpiece interface and the friction coefficient as well [17, 19, 20]. Moreover, it was reported in that [17] the coefficient is highly influenced by the axial force, the friction force, the geometry of the contact area, the welding parameters and the properties of the material. In the meantime, as observed from Fig. 4, when the temperature increases more than 120 °C, the material strength decreases resulting in the decrease of heat generation and the friction coefficient. Schmidt and Hattel [21] introduced the concept of the self-stabilizing effect, which was accountable whenever the temperature level is below the solidus temperature. It was also observed that when the temperature is higher than the solidus temperature, the material yield stress decreases. In contrast, as the temperature decrease, the material yield stress recovers [22]. This observation supports the reduction of the friction coefficient values by decreasing of the temperature. Aziz et al. [23] and Schmidt and Hattel [24] developed heat generation models in which the friction coefficient reduced with the rise of the temperature and descended almost to zero as the temperature is increased to the material melting point. Tao et al. [17] used the material removal tests, and discovered that the friction coefficient during the welding is temperature dependent. It was observed that, the friction coefficient was around 1.1 at room temperature and 1.3 at 250 °C, then, it reduced with the rise of the temperature. The study also reported that, as the temperature increased from the room temperature (25 °C) to 250 °C, the cohesion between two friction bodies increased to the maximum value of the cohesion. The report indicated that when the temperature is around 250 °C and beyond, the strengthening β'' phase of the AA6061Al-T6 will transform into the α phase. Since the strength of the material in the α phase is lower than the final strength of the material, reduce in the friction coefficient occurred. It should be noted that the friction coefficient values which is found in Tao et al.'s study [17], is different from those in the present and other studies. This is probably due to the various experimental conditions. In a study by Zhang et al. it was reported that [17] the friction coefficient rose to a maximum value of 0.4 at 200 °C. However, the values began to decrease when the temperature increased to more than 200 °C. These facts suggest that the friction coefficient exhibits a strong correlation with temperature. In addition, as can be shown in Fig. 4, the variation of the friction coefficient and the temperature is approximately described as a piecewise normal function when the temperature is higher than 120 °C.

It is also observed that the value of the friction coefficient falls very close to zero when the welding temperature is close to the material melting point. As mentioned earlier, during the FSW, the welding temperature varies in a range of 80–90% of the base material melting point [18], therefore, the calculated values of the friction coefficient are applicable for all welding steps such as plunging, dwelling, stirring, and retraction.

5.2 Temperature Distribution Results at Different Welding Stages

Figure 5 shows the temperature distribution during different welding steps. As can be seen, transient phases are present in the plunging, dwelling and welding steps until the steady state condition reaches. Due to the use of the ALE technique the mesh is gradually refined towards the welding path.

In the first step (plunging step), for achieving the sufficient temperature, a stationary rotating tool plunges into the workpiece. It is observed that, during the plunging and the dwelling steps, the formation of the burr behind the tool is higher in comparison with the front side, because the temperature in the back side of the tool is higher. In addition, the highest temperature is extracted from the dwelling step (around 601 °C). As reported by researchers, during the dwelling step, the axial force and the torque applied to the tool are higher than other steps [25]. To illustrate, at the beginning of the welding higher loads are applied in order to stir and deform the material, because the temperature is not sufficiently enough to deform the material.

The results show that in the welding step the temperature is around 585 °C, hence as the material gets softer the force and the torque which are applied to the tool are also decreased. As the tool begins to move along the welding path, the temperature around the tool changes in an unpredictable way, thereby there is a transient period before the necessary steady-state is reached. Then, the peak temperature after the dwelling step fell slightly. In addition, in the welding step a pseudo steady-state status has appeared. It should be noted that, even though the heat generation fluctuate, the thermal field remains effectively constant around the tool. Furthermore, in the welding step from the shoulder downward, the flow line exhibits a small change over the depth. It is also observed that; the heat reflects from the end of the plate causing extra heating around the tool.

5.3 Temperature Profile at the Welding Cross Section

Figure 6 shows that, the temperature profile below the shoulder is asymmetrically distributed due to the non-uniform material flow behaviour near the welding tool and also the higher deposit of material in the advancing side. Moreover, in both

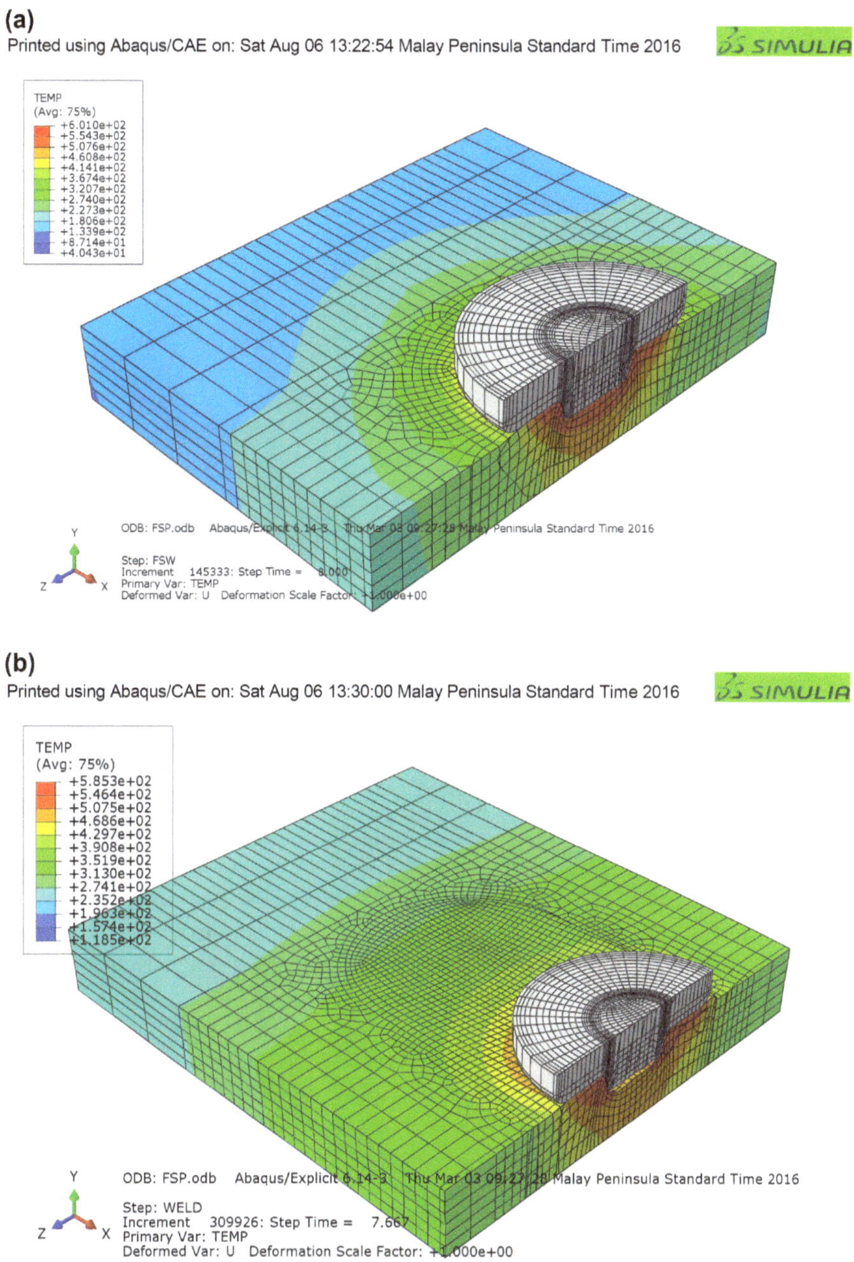

Fig. 5 The temperature field in different steps of the welding **a** plunging **b** welding **c** end of the welding

(c)

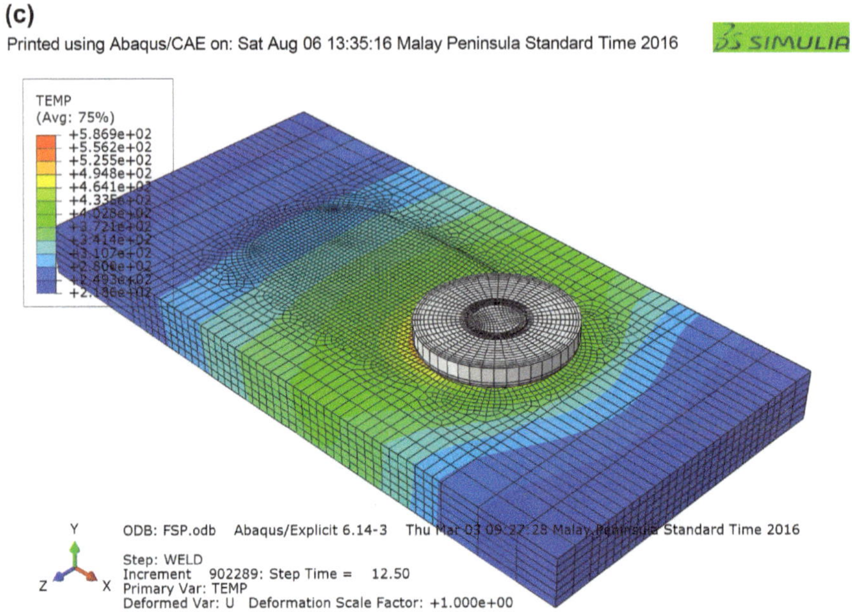

Fig. 5 (continued)

experimental and numerical results, it is found that, the temperature has an upward trend pattern during the welding in both sides of the welding (advancing and retreating sides), but started to give a distinguished temperature differences on each side within 10 s approximately. To illustrate, the strain rate and the temperature in the advancing side is higher in comparison with the retreating side (approximately about 30 °C), because in the advancing side, the velocity component is positive resulting more shearing action of the material in comparison to the retreating side. To illustrate, since the welding tool transverse movement resist against the shoulder and the pin rotational movement, the temperature in the advancing side is higher in comparison with the retreating side. Meanwhile, the difference in the velocity gradient (the positive component of the velocity) on both advancing and retreating sides of the weld makes this asymmetrical balance in the HAZ area. Additionally, since the higher stirring on the top surface caused higher deformation of the material, the effective strain of the work-piece is decreased from the top surface of the workpiece to the bottom surface. Furthermore, higher strain is observed at the welding centre line in comparison with the other parts, because the rotational movement of the pin mainly shapes the stirred zone. Besides, it should be considered that when the welding transverse velocity is constant, with the increase of the rotational speed, the stirring zone size is also increased, because of the effective material mixing. In contrast, with the rise of the transverse velocity when the welding rotational velocity is constant, the stirring zone size is decreased. Therefore, at higher transverse speeds the material exhibits relatively more

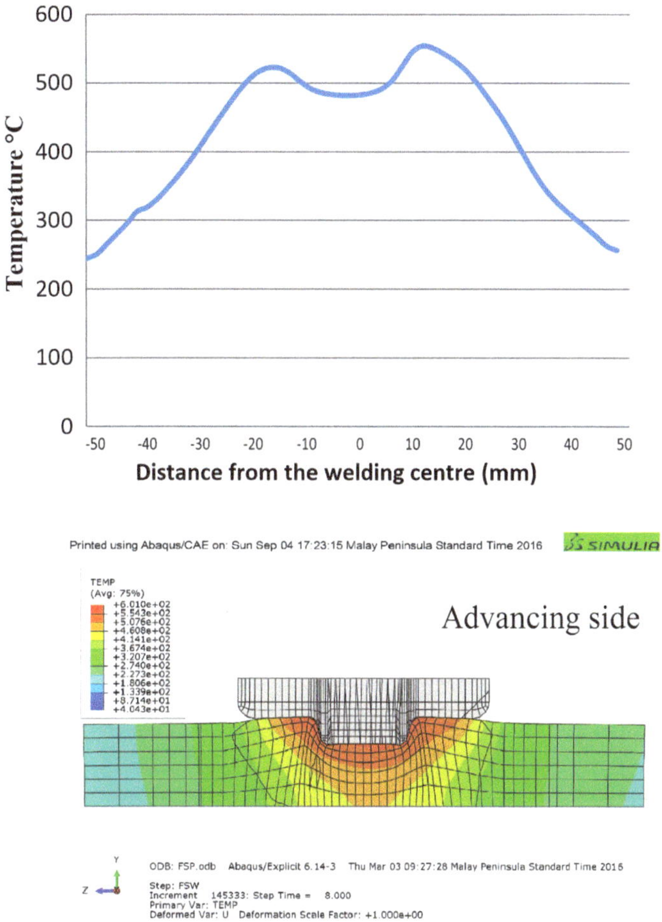

Fig. 6 Temperature fields and the temperature distribution across the welding cross section

resistance to the tool transverse movement and the material will not mix enough to achieve a high-quality weld. Furthermore, it is observed that 10 mm away from the welding centre line, the temperature is higher in comparison with the welding centre line (about 70 °C in the advancing side and about 50 °C in the retreating side). This happens due to the higher friction force at the pin corner compared with the lower pin surface.

5.4 Temperature Distribution in the Top Surface

Figure 7 illustrates that, the temperature pattern on the top surface has not changed significantly and the temperature tends to follow a coincident pattern, thus at this

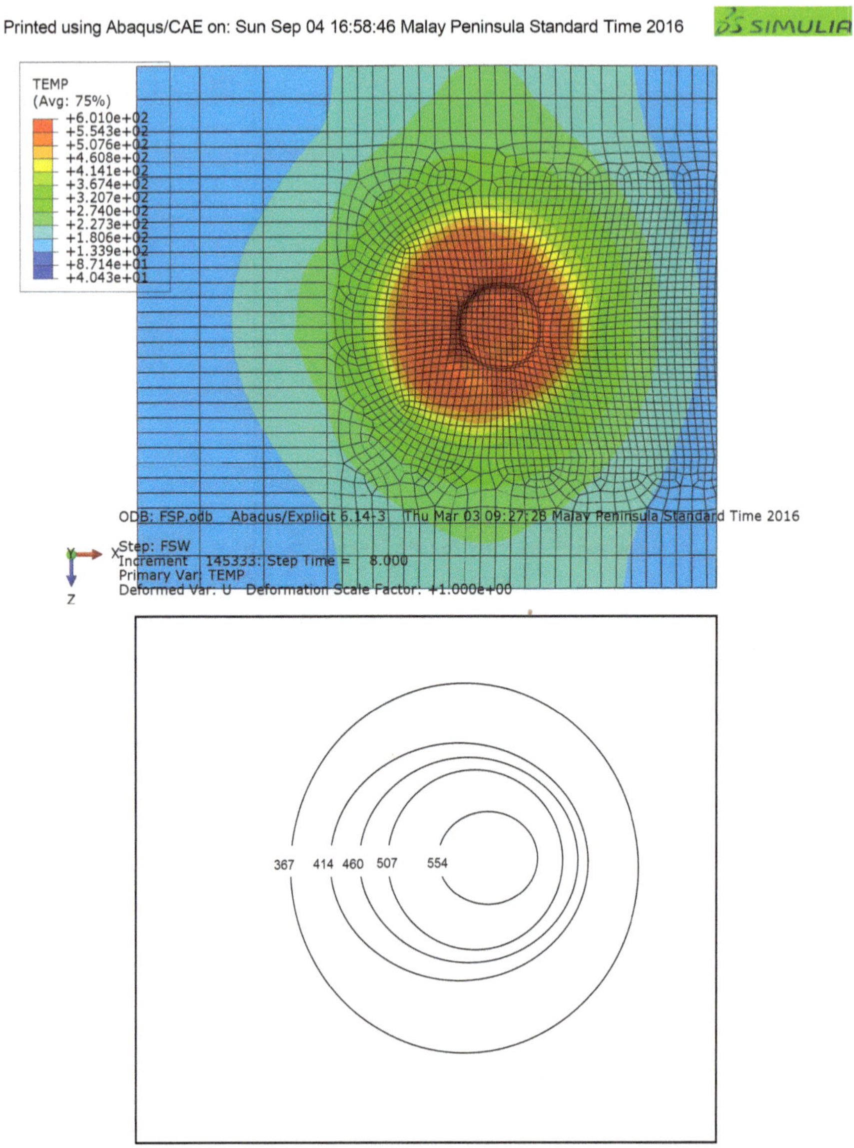

Fig. 7 The temperature distribution at the top surface

surface the heat transfer can be assumed as a quasi-steady condition. Moreover, in the HAZ area, the temperature pattern is skewed toward the advancing side and the leading edge with the highest temperature of 509 °C (takes place around the shoulder area). Consequently, as the tool moves along the welding path, the cooler material (the material that has not been welded yet) which is located in the leading edge (tool front side), is swept to the advancing side. After stirring of the material by the tool, it moves from the advancing side to the trailing edge (tool back side) and it is deposited to the advancing side or it is flowed toward the workpiece thickness. This issue makes the temperature of the trailing edge and the advancing side higher in comparison with the leading-edge and the retreating side, respectively.

5.5 *Temperature Comparison of Different Friction Coefficient Values*

A comparison between two different numerical models and experimental observations in the field of the highest measured temperate is shown in Fig. 8.

In the first numerical model, the optimum value of the friction coefficient, which is documented by the literature (0.3) is used [14, 15, 24, 26]. It is observed that, when the documented value is used in different models, the job could not continue after the plunging step, and it is aborted due to the excessive distortion of mesh. Therefore, the model only could simulate the plunging step. Meanwhile, as shown in Fig. 8, the predicted values after the plunging step are not in a good accordance with the experiments.

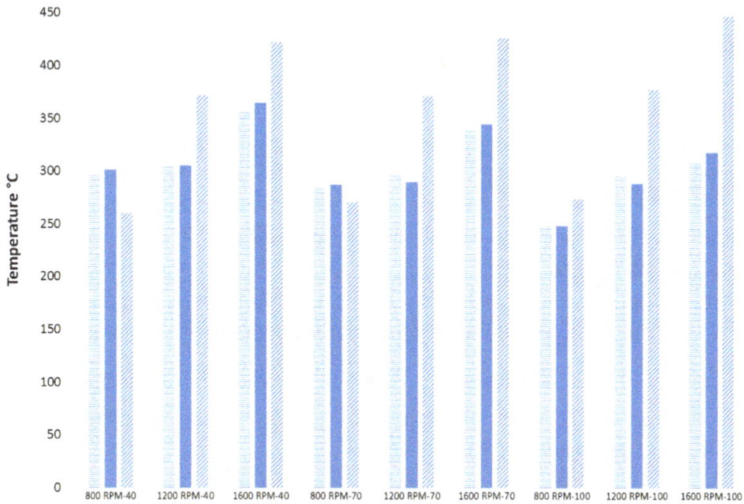

	800 RPM-40 MM/Min	1200 RPM-40 MM/Min	1600 RPM-40 MM/Min	800 RPM-70 MM/Min	1200 RPM-70 MM/Min	1600 RPM-70 MM/Min	800 RPM-100 MM/Min	1200 RPM-100 MM/Min	1600 RPM-100 MM/Min
Experiment (°C)	295.828	304.592	357.146	285.366	295.828	338.173	247.956	296.551	308.893
present study (°C)	301.267	305.529	365.044	287.57	290.034	344.461	248.565	288.301	318.005
Literature values [14, 15, 25, 27] (°C)	260.305	372.416	422.59	270.692	371.09	426.522	274.328	377.716	446.892

Experiment (°C) ■ present study (°C) Literature values [14, 15, 25, 27] (°C)

Fig. 8 Numerical and experimental results for the peak temperatures in different rotational (RPM) and transverse (mm/min) speeds

In the second numerical model, the calculated values of the temperature dependent friction coefficient values (from 0.207089 to 0.00058) which are measured by the mathematical formulation are employed. It is observed that; the simulation could continue by the end of the welding (plunging-out step). Moreover, in the term of the temperature, good agreements between the results of the developed FEM (by the mathematical model) and the experimental measurements have been found and the gap between the experimental measurements and the predicted temperature values is decreased.

In addition, many authors [20, 27–29] claimed that, the heat generation strongly depends on the welding speeds (i.e. rotational and transverse speeds).

The results illustrate that, when the welding transverse speed is 40 mm/min the highest measured temperature in the rotational speed of 800 RPM is equals to 301 °C, but when the welding rotational velocity increases to 1200 RPM and 1600 RPM, a significant increase in the temperature field is observed, 305 °C and 365 °C, respectively. This rise is because of the higher stirring of the material at higher rotational speeds. To illustrate, when the stirring phenomena increase, the material becomes more ductile, thus the welding force, the material resistance and the contact pressure decrease sharply.

Furthermore, the material mixing at the lower half part of the nugget zone is higher when higher rotational velocities are used. This well-done mixing of the material improves the joining quality. Similar results have been reported in other literature [27–29].

In addition, when the welding transverse speed is 40 mm/min and the welding rotational speed is 800 RPM, the maximum temperature equals to 301 °C, however, when the transverse speed increases from 40 to 100 mm/min the peak temperature drops from 301 to 248 °C. Thus, the highest measured temperature drops with the rise of the welding transverse velocity. This decrease happens due to the heat dissipates over a wider region of the workpiece and the reduction of the input heat per unit length during higher transverse speeds. Therefore, it is observed that, the material is comparatively harder at higher transverse speeds, leading to the increase of the forces acting on the tool which affects the tool life [14, 29]. In contrast, when the transverse speed is relatively low, the material which is located in the front of the tool is exposed to higher temperature, therefore the material becomes softer and the forces that are applied to the tool decrease [14, 29].

6 Conclusions

In this paper, a mathematical formulation was used to predict temperature dependent values of the friction coefficient. Then, the calculated values were employed to increase the accuracy of a three-dimensional FEM and also to solve the divergence phenomenon that takes place after the welding plunging step. The results of the FEM showed that, the heat which is generated by the shoulder is higher than the heat which is produced by the pin (approximately around 60 °C difference).

Moreover, below the shoulder, the temperature profile followed an asymmetrical pattern. In addition, the temperature at the advancing side was observed about 30 °C higher compared to the retreating side. It was also detected that, there is a small gap (approximately around 10–15 °C) between the upper and the lower surfaces in the stirring zone. Additionally, the peak temperature variations were detected as a quasi-steady phenomenon in the stirring zone, however the quasi-steady phenomenon faded, as the distance from the welding path increased. It was also found that, by increasing the rotational velocity in a constant transverse speed, the peak temperature at the workpiece gradually increases. Meanwhile, unlike the rotational speed, a decrease in the temperature values was observed at higher transverse speeds. Finally, the FEM results were verified by experimental observations and published papers.

References

1. Widener C et al (2006) High-rotational speed friction stir welding with a fixed shoulder. In: Proceedings of the 6th international symposium on friction stir welding, Montreal, Canada
2. de Saracibar CA et al (2014) Computational modeling and sub-grid scale stabilization of incompressibility and convection in the numerical simulation of friction stir welding processes. Archives of Computational Methods in Engineering 21(1):3–37
3. Assidi M, Fourment L (2009) Accurate 3D friction stir welding simulation tool based on friction model calibration. Int J Mater Form 2(1):327–330
4. Deng X, Xu S (2004) Two-dimensional finite element simulation of material flow in the friction stir welding process. J Manuf Process 6(2):125–133
5. Schmidt HNB, Hattel J (2004) Modelling thermomechanical conditions at the tool/matrix interface in friction stir welding. In Proceedings of 5th international friction stir welding symposium. The Welding Institute
6. Santiago D et al (2010) Modelado termomecánico del proceso de Friction Stir Welding utilizando la geometría real de la herramienta. Revista internacional de métodos numéricos para cálculo y diseño en ingeniería 26(4):293–303
7. Chiumenti M et al (2013) Numerical modeling of friction stir welding processes. Comput Methods Appl Mech Eng 254:353–369
8. Bussetta P et al (2014) Comparison of a fluid and a solid approach for the numerical simulation of Friction Stir Welding with a non-cylindrical pin. Steel Res Int 85(6):968–979
9. Bussetta P et al (2015) 3D numerical models of FSW processes with non-cylindrical pin. Adv Mater Process Technol 1(3–4):275–287
10. Dialami N et al (2013) An apropos kinematic framework for the numerical modeling of friction stir welding. Comput Struct 117:48–57
11. Dialami N et al (2015) Material flow visualization in friction stir welding via particle tracing. Int J Mater Form 8(2):167–181
12. Dialami N et al (2014) Numerical simulation and visualization of material flow in friction stir welding via particle tracing. In: Numerical simulations of coupled problems in engineering. Springer, pp 157–169
13. Chiumenti M et al (2013) A novel stress-accurate FE technology for highly non-linear analysis with incompressibility constraint. Application to the numerical simulation of the FSW process. In: AIP conference proceedings. Proceedings of the international conference on numerical methods in forming processes, NUMIFORM

14. Chao YJ, Qi X, Tang W (2003) Heat transfer in friction stir welding—experimental and numerical studies. J Manuf Sci Eng 125(1):138–145
15. Chen C, Kovacevic R (2003) Finite element modeling of friction stir welding—thermal and thermomechanical analysis. Int J Mach Tools Manuf 43(13):1319–1326
16. Guide FD (1988) An engineering study of coefficient of friction of materials and coatings. Slide-chart presentation from General Magnaplate Corp, Linden
17. Zhang X, Xiao B, Ma Z (2011) A transient thermal model for friction stir weld. Part I: The model. Metall Mater Trans A 42(10):3218–3228
18. Tang W et al (1998) Heat input and temperature distribution in friction stir welding. J Mater Process Manuf Sci 7:163–172
19. Su H et al (2014) Thermal energy generation and distribution in friction stir welding of aluminum alloys. Energy 77:720–731
20. Su H et al (2015) Numerical modeling for the effect of pin profiles on thermal and material flow characteristics in friction stir welding. Mater Des 77:114–125
21. Zhang X, Xiao B, Ma Z (2011) A transient thermal model for friction stir weld. Part II: Effects of weld conditions. Metall Mater Trans A 42(10):3229–3239
22. Zhang Z (2008) Comparison of two contact models in the simulation of friction stir welding process. J Mater Sci 43(17):5867–5877
23. Aziz SB et al (2016) Impact of Friction Stir Welding (FSW) process parameters on thermal modeling and heat generation of aluminum alloy joints. Acta Metall Sin (English Lett) 29 (9):869–883
24. Schmidt HB, Hattel JH (2008) Thermal modelling of friction stir welding. Scripta Mater 58 (5):332–337
25. Fraser K, St-Georges L, Kiss LI (2016) A mesh-free solid-mechanics approach for simulating the friction stir-welding process
26. Song M, Kovacevic R (2003) Thermal modeling of friction stir welding in a moving coordinate system and its validation. Int J Mach Tools Manuf 43(6):605–615
27. Khodir SA, Shibayanagi T, Naka M (2006) Microstructure and mechanical properties of friction stir welded AA2024-T3 aluminum alloy. Mater Trans 47(1):185–193
28. Colegrove P, Shercliff H (2003) Experimental and numerical analysis of aluminium alloy 7075-T7351 friction stir welds. Sci Technol Weld Joining 8(5):360–368
29. Padmanaban R, Kishore VR, Balusamy V (2014) Numerical simulation of temperature distribution and material flow during friction stir welding of dissimilar aluminum alloys. Proc Eng 97:854–863

FSC
www.fsc.org
MIX
Papier | Fördert
gute Waldnutzung
FSC® C083411

Kolibri 360 GmbH
Ferdinand-Jühlke-Straße 7
99095 Erfurt, Deutschland
produktsicherheit@kolibri360.de